# MEASURING RECREATION SUPPLY

WINSTON HARRINGTON

A STUDY FROM **ResOURCes**
**FOR THE FUTURE**

Washington, D.C.

Printed in the United States of America

Published by Resources for the Future
1616 P Street, N.W., Washington, D.C. 20036

Books from Resources for the Future are distributed worldwide by The Johns Hopkins University Press

**Library of Congress Cataloging-in-Publication Data**

Harrington, Winston.
  Measuring recreation supply.

  Bibliography: p.
  1. Outdoor recreation—Economic aspects.  2. Nonrenewable natural resources—Economic aspects.  3. Outdoor recreation—Government policy.  I. Resources for the Future.  II. Title.
  GV191.6.H37  1987        333.78'11        87-9569
  ISBN 0-915707-31-4 (alk. paper)

⊗ The paper in this book meets the guidelines for permanence and durability of the Committee on Production Guidelines for Book Longevity of the Council on Library Resources.

**Winston Harrington** is a Fellow in the Quality of the Environment Division at Resources for the Future.

RESOURCES FOR THE FUTURE (RFF) is an independent nonprofit organization that advances research and public education in the development, conservation, and use of natural resources and in the quality of the environment. Established in 1952 with the cooperation of the Ford Foundation, it is supported by an endowment and by grants from foundations, government agencies, and corporations. Grants are accepted on the condition that RFF is solely responsible for the conduct of its research and the dissemination of its work to the public. The organization does not perform proprietary research.

RFF research is primarily social scientific, especially economic, and is concerned with the relationship of people to the natural environment—the basic resources of land, water, and air; the products and services derived from them; and the effects of production and consumption on environmental quality and human health and well-being. Grouped into three research divisions—Energy and Materials, Quality of the Environment, and Renewable Resources—staff members pursue a wide variety of interests, including food and agricultural policy, water supply and allocation, forest economics, risk management, natural gas policy, multiple use of public lands, mineral economics, air and water pollution, energy and national security, hazardous wastes, climate resources, and the economics of outer space. Resident staff members conduct most of the organization's work; a few others carry out research elsewhere under grants from RFF.

Resources for the Future takes responsibility for the selection of subjects for study and for the appointment of fellows, as well as for their freedom of inquiry. The views of RFF staff members and the interpretations and conclusions of RFF publications should not be attributed to Resources for the Future, its directors, or its officers. As an organization, RFF does not take positions on laws, policies, or events, nor does it lobby.

# CONTENTS

# PREFACE

This study was begun early in 1985, at about the same time that a national recreation commission was established to examine the current status and future prospects of outdoor recreation in the United States. One of the more prominent voices calling for the establishment of such a commission was the Outdoor Recreation Policy Review Group, chaired by Henry Diamond and headquartered at Resources for the Future.

The President's Commission on Americans Outdoors has now submitted a report to the President.[1] The principal recommendation of the commission calls for the contribution of at least a billion dollars a year to a trust fund dedicated to improvements in outdoor recreation opportunities. This controversial proposal has drawn the fire of both fiscal conservatives, who decry the vast commitment of resources for this purpose, and environmental groups and other advocates of recreation opportunities, who believe it to be inadequate to meet future recreation needs.

The controversy reflects the great difference in the values and preferences of the disputants. But it is also a reflection of the fact that we have no way of characterizing the availability of recreation resources, except perhaps in the context of a single recreation site. At the macrolevel, we cannot in any meaningful way compare recreation opportunities in different geographic regions, or over time in the same region. Nor can we say how those opportunities change if new recreation areas are opened or existing ones expanded. And without

---

[1]At this writing, the final report of the commission is not yet available to the public, although a summary publication, "Americans and the Outdoors," has been distributed by the commission.

being able to say how much recreation we have, how can we possibly say how much we *should* have?

This study is an attempt to define recreation resource availability so as to make such comparisons meaningful. The approach used here is borrowed from the economics of nonrenewable natural resources, where resource availability (or more commonly its inverse, scarcity) is the central issue. The leading candidates for useful measures of the scarcity of a natural resource are various indicators of its price or cost, such as unit extraction cost or product price. These indicators share the property that they summarize, to a greater or lesser degree depending on the resource and the context, the sacrifices that must be endured to obtain a unit of the resource. Similarly, knowing the availability of recreation resources requires an answer to the following question: How much effort is required for an individual or household to enjoy a recreation experience? With recreation, however, the price of entry to a recreation site is usually only a small part of the sacrifices that must be endured. To develop a recreation scarcity measure, these other sacrifices must somehow be quantified and converted to a common metric.

The problem of characterizing recreation supply was posed to me by Emery Castle, former president of RFF and member of the Outdoor Recreation Policy Review Group. I am grateful to him and to Clifford S. Russell, former director of the Quality of the Environment Division at RFF, for their support and encouragement and for providing thoughtful comments on an early draft. Helpful comments were also provided by Marion Clawson, John V. Krutilla, Wallace E. Oates, Paul Portney, and especially V. Kerry Smith, who entered into an extensive written correspondence with me regarding the issues raised in the study. Anonymous reviewers of the manuscript also provided unusually helpful comments. As always, they are absolved from any responsibility for any errors and omissions that may remain.

I am also grateful for the excellent work of Cheryl Johnson, who typed the various drafts of the manuscript, and Martha Bari, production editor. Finally, I would like to thank Barbara Karni, a very capable editor.

*December 1986*                                        *Winston Harrington one*

# THE PROBLEM OF RECREATION SUPPLY

On January 28, 1985, President Reagan signed an executive order establishing a new Presidential Commission on Outdoor Recreation Resources Review. The purpose of the commission is to examine outdoor recreation policies and opportunities provided by all levels of governments and by the private sector, and to "make recommendations to the President concerning the outdoor recreation resources, programs, and opportunities that will insure the future availability of outdoor recreation for the American people."[1] This commission is patterned after the Outdoor Recreation Resources Review Commission (ORRRC), a highly influential commission established by Congress in 1958. In setting the course of U.S. recreation policy for the next quarter of a century, the ORRRC was responsible for many ideas that have since become integral parts of the recreation establishment, including wilderness, wild and scenic rivers, and the Land and Water Conservation Fund.

The impetus for the new commission came, to some degree, from the Outdoor Recreation Policy Review Group, a study group sponsored by Laurence Rockefeller and chaired by Henry L. Diamond. This group believed that another long look at U.S. recreation policies was called for in view of the substantial changes in the structure of American life that had occurred since 1960. These changes included the new status of women, migration into the Sunbelt, and the aging of the population. In addition, most of the group's members viewed with alarm evidence of a decline in the quantity and quality of recreation resources (Diamond and coauthors, 1983).[2] This evidence in-

---

[1]Executive Order, "Presidential Commission on Outdoor Recreation Resources Review," January 28, 1985.

[2]For convenience, this study will be referred to as the Diamond Report.

cluded a decline in the recreation acreage per capita in most regions of the country, the continued regional imbalance between East and West in recreation availability, and the growing limitations on access to recreational lands. In particular, the group noted an 8 percent decline in the number of national forest campgrounds and picnic areas and the closure or partial closure of over four hundred recreation sites maintained by the U.S. Army Corps of Engineers. Some evidence also suggests that the quality of facilities is deteriorating at many national parks.[3] This deterioration has mainly been the result of the cut in spending, in an era of chronic budget deficits. For example, between 1960 and 1980 expenditures by the National Park Service increased at an average annual rate of 5.4 percent, while visits at facilities administered by the Park Service increased at a rate of 6.7 percent. Between 1981 and 1984, expenditures *declined* by 6.3 percent per year, with annual visits remaining constant. Similarly, total expenditures and visits at state parks increased at average annual rates of 9.0 and 3.9 percent, respectively, between 1960 and 1979. Expenditures declined by 11 percent per year between 1979 and 1984, while total visits were increasing by 5.6 percent per year.

This set of facts led the Diamond Report to conclude that "governments in general are doing less for outdoor recreation than is required to meet the need" (p. 26). That is, the demand for recreation resources is increasing faster than the supply.

Fears of looming resource scarcity are a recurring phenomenon in resource economics. In the early 1800s, Malthus and Ricardo worried about the coming shortage of agricultural land; in the 1870s, Jevons anticipated a crisis in coal; in the early years of this century, U.S. concerns about timber and petroleum shortages led to the establishment of the National Forest System and the Naval Petroleum Reserves at Elk Hills and Teapot Dome. During the 1970s, concern over nearly all resources heightened, partly as a result of the actions of OPEC and the findings of the first Club of Rome report. For marketed resources these fears have now vanished, at least temporarily, amid a worldwide commodities glut. For marketed commodities, shortages were avoided by discoveries of new sources of supply or new technology, induced mainly by the reality or prospect of higher prices.

Concern remains regarding nonmarketed resources, such as water, air, the ozone layer, and endangered species (see Barney, 1979, for example). Since they are for the most part provided by governments on public land, outdoor recreation services number among the latter type of resource. The possibility of a growing imbalance between recreation supply and demand is a matter of some policy concern. Determining the adequacy of recreation supply is difficult, however.

[3]This trend may be reversed by the recent efforts by the Department of Interior to improve conditions (GAO, 1980, 1982).

In part, the problem is empirical. Although there is a mass of data on both recreation participation and resources, the data are largely unassembled and often incommensurate. But by far the greatest difficulties are conceptual, especially on the supply side. Discussion of recreation policy is hampered by the lack of agreement on what the supply of recreation is and how it should be measured. Without adequate supply measures, it is difficult to discuss the key issues associated with recreation policy in the aggregate. How are recreation resources distributed within and among the various regions of the country? How is the availability of recreation resources changing over time? Where and when should new recreation facilities be built, and how should they be managed?

This study is an attempt to formulate a more precise and economically meaningful notion of recreation supply and recreation resource scarcity (or its complement, availability). This will be largely a theoretical exercise, though some attention will be paid to the kinds of data and empirical analyses that would be necessary to measure the concepts discussed here. The aim is to develop a notion of resource scarcity comparable to that used in the analysis of marketable natural resources, such as minerals or timber. A measure of resource scarcity has but one essential property: "it should summarize the sacrifices, direct and indirect, made to obtain a unit of the resource" (Fisher, 1979).

For a conventional marketed commodity, this property is best conveyed by the good's price. As the price of the good increases, by definition, it becomes more scarce. The price of a good—and hence its scarcity—is determined by the intersection of the supply and demand curves. An upward shift in the demand or supply curve causes an increase in the scarcity of the good. Only when supply is perfectly elastic is scarcity unaffected by an increase in demand.

In practice, the appropriate indicator of scarcity is often less clear. What, for example, is the best indicator of the scarcity of petroleum? Is it the price of crude oil? The cost of extraction? Or, perhaps, the price of gasoline and other petroleum products? During the 1970s, the world price of crude and the marginal cost of extraction diverged greatly, owing to the market power of OPEC. This contrived scarcity encouraged both conservation and new, higher-cost production in non-OPEC countries. In the face of these trends, the artificially high OPEC price could not be sustained and by 1986 had fallen almost to pre-1972 levels.

Scarcity of finished petroleum products was manifested differently. The U.S. prices of finished products were held down by price controls. But even though gasoline prices remained relatively low, scarcity was not always avoided, as in 1974 and 1979, when long lines appeared at gasoline stations. In these circumstances, the nominal price of gasoline was a relatively small part of the sacrifice necessary to obtain a tankful of gasoline.

The implications of this story for measuring scarcity are, first, that several relevant prices may provide conflicting signals about scarcity. Second, prices are subject to distortion as a result of market power or market intervention, and may mislead thereby. Third, there is a difference between the short and long run, and reaction to short-run scarcity often leads to its elimination. Fourth, the relevant "price" is sometimes different from the money price, as when consumers endure increased waiting time or uncertainty.

The appropriateness of various scarcity measures remains somewhat of an open issue. For example, in *Scarcity and Growth* (Barnett and Morse, 1963)—one of the most notable attempts to measure resource scarcity—the authors proposed as a scarcity measure the unit extraction cost, and used this measure to test hypotheses of increasing resource scarcity by examining cost trends for minerals and timber for the past century or so. In a retrospective on this work (Smith, 1979), one of the main issues was whether extraction cost was the appropriate measure of scarcity, other candidates being the owner's economic rent and the prices of products employing such resources. In short, defining scarcity is simple in the abstract but less so in practice, particularly for nonmarketed goods, such as recreation.

The measurement of recreation scarcity is particularly problematic because of the misleading nature of physical measures of recreation supply; because of the role that consumers play in providing inputs as well as consuming outputs; and because of the nonpecuniary nature of costs. Ordinarily, recreation supply is measured in terms of physical units, such as acres, miles of hiking trails, or number of campsites. Such measures are misleading, however, because they fail to take into account how people use recreational facilities, quantifying instead merely the physical measure of such resources. Recreational participants derive satisfaction not from acres of land but from the activities they pursue at a facility (for example, from camping, hiking, swimming). The "outputs" of different recreations areas, though typically measured in common units of participation, such as visits, visitor days, or participation days, are not homogenous. A myriad of activities fall under the heading of "outdoor recreation." Even among activities of the same kind, important quality differences may exist. (Compare, for example, the value of a day skiing at Aspen and a day skiing at the local county park.) This makes it difficult to make comparisons of recreation experiences at different sites or of different types. To be sure, heterogeneity of output exists among many goods, but what we term "recreation" is heterogeneous in the extreme.[4]

[4]This great heterogeneity represents an aggregation problem, which appears in many forms in economics. The essence of the problem is the need to determine what general

4

But measuring outputs in terms of recreational experiences leads to a second problem. In the production of recreation outputs, inputs are provided not only by the site operator but by the recreation participant as well. This is very different from conventional production theory, where producers produce and consumers consume.

Finally, recreation is unusual because the sacrifices that must be endured to enjoy an experience are not necessarily, or even primarily, pecuniary. Above it was noted that nonmonetary costs can arise even for market goods; for nonmarket goods they normally play a much larger role. As Baden and Stankey (1975) point out, people "pay" for recreation experiences in several ways: by devoting time to the experience, by waiting in line, by enduring diminished quality at congested sites, and perhaps by being excluded from some sites.

Since such valuations are not observed in market transactions, they must be inferred from a variety of methods, the development of which has occupied applied welfare economists for the past quarter of a century. A scarcity indicator for coal, for example, should include the cost of the environmental damage associated with its extraction, something not reflected in the price of coal. Needless to say, less confidence can be placed in such valuations than in market prices.

In spite of these difficulties, it will be argued here that recreation resource scarcity can usefully be approached in a manner analogous to that used for market commodities, by converting the costs incurred by both recreation participants and the operators of recreation areas into a common metric that will serve as a scarcity measure. Just as with market commodities, several price-related concepts can be defined, depending on the uses at hand. In what follows, we shall deal with two such measures. The first is a relationship between the social cost of producing a recreation experience and the number of participants served. This will correspond roughly to short- or long-run cost curves from the theory of production. Second, we shall define a price of a recreation experience that corresponds to the price of final products. It turns out that this price—which will be called an "effective price," to emphasize that more than money prices are involved—is specific to the residential location of the user and the type of recreation activity pursued. The nonmonetary costs of greatest interest are travel cost and the cost of congestion, both of which already figure prominently in recreation economics. However, their roles here will be a bit unconventional.

---

statements can be made about a collection of disparate objects. That is, even though individual recreation areas all differ, we need to say something about recreation areas in the aggregate in order to design policies at the aggregate level. It should be noted that the current measures of recreation supply, concerned mainly with the physical quantity of recreation resources, such as area of parkland, miles of shoreline or hiking trails, and the like, also suffer from problems of aggregation.

## Plan of the Study

Existing measures of recreation resource availability are reviewed briefly in chapter 2. In chapter 3, production and cost functions for a single recreation site will be defined, and it will be shown how these functions depend on how the site is managed. In chapter 4 we consider the aggregate of recreation resources in a region and use the results of the previous chapter to define a "price" of a recreation experience to a participant in the region. This price is a measure of the availability of recreation resources to that participant. In chapter 5 we discuss in some detail one of the most important elements of this price, the cost of congestion. We conclude (chapter 6) with another look at the major empirical problems likely to be encountered in devising recreation scarcity measures, the additional data that are needed, and what, if anything, can be done with existing data.

*two*

# EXISTING MEASURES OF SUPPLY

Recreation resources are usually measured in physical terms, especially in terms of available acreage. The most comprehensive summary of outdoor recreation data is found in Clawson and Van Doren (1984), who list acreage—and in some cases other physical measures, such as trail mileage—of recreation resources administered by the various federal agencies and state governments. More detailed data can be found in more specialized publications. The National Association of State Park Directors (1984) provides an inventory of acreage, trail mileage, cabins, campsites, and so forth for each state park system. The Statewide Comprehensive Outdoor Recreation Plans (SCORPs) often list still more detailed information, such as number of picnic tables, fire pits, ball fields, and miles of shoreline, often for substate regions or even individual parks.

Physical measures of market resources are frequently put forward as indicators of scarcity. Thus, for energy and mineral resources, we find estimates of total resources, recoverable reserves, and the like. The usefulness of such indicators is limited by their disregard for the quality of the deposits, the cost of extraction, and the distance to markets. Physical measures of recreation resources can be similarly misleading or uninformative as indicators of resource availability. To see this, let us consider the spatial distribution of recreation resources in the United States. The great majority of land available for outdoor recreation in the United States is located in the western half of the country. Since the majority of the population is located in the eastern states, this imbalance is even more pronounced when expressed in terms of the acreage of recreational land per capita. This distribution of recreational land has suggested to many observers that recreational opportunities are similarly maldistributed, and hence that additional recreational land is needed in the more populated eastern half of the country.

In support of this view, the Diamond Report (mentioned in chapter 1) included two tables, reproduced here as tables 2–1 and 2–2. Table 2–1 compares per capita recreation land in the various sections of the country in the 1960s and the 1980s. It shows that acreage per capita available for recreation in 1980 ranged from 0.10 in the Mid-Atlantic states to 9.65 in the Rocky Mountain states, so that the latter apparently enjoyed about 100 times the area per person as the former. The general pattern is one of greater recreation availability as one moves westward. Except for the Mountain and Pacific states, where a large decline was experienced because of the large increase in population, these figures are approximately the same as they were in 1960. To determine whether this East–West imbalance has been improved over the last two decades, the Diamond Report examined trends in acreage per capita with population held fixed. As shown in table 2–2, all but one of the regions in the western half of the country had changes in acreage per capita greater than the national average, while none of the eastern regions did so. In general, those regions with the largest per capita acreage to start with showed the most improvement, if population is held constant. Thus, the Diamond Report concluded not only that a large regional imbalance in recreation acreage exists, but also that "the disparity in the supply of recreation land between West and East has not improved significantly since 1960."

However, when different data are examined, or when these same data are examined in a different light, a different and much less compelling picture comes into view. Table 2–3 shows the regional

**Table 2–1. Per Capita Distribution of Public Recreation Land, 1960 and 1980**

(recreation acres per capita)

| Region | 1960 | 1980 | Change |
|---|---|---|---|
| New England | 0.17 | 0.16 | −0.01 |
| Mid Atlantic | 0.11 | 0.10 | −0.01 |
| East North Central | 0.16 | 0.17 | +0.01 |
| West North Central | 0.45 | 0.52 | +0.07 |
| South Atlantic | 0.36 | 0.21 | −0.15 |
| East South Central | 0.35 | 0.31 | −0.02 |
| West South Central | 0.32 | 0.27 | −0.05 |
| Mountain | 15.59 | 9.65 | −5.94 |
| Pacific[a] | 2.45 | 1.71 | −0.74 |
| United States | 1.09 | 0.92 | −0.17 |

Note: Includes only areas specifically designated for parks and recreation.
Source: Henry L. Diamond and coauthors, "Outdoor Recreation for America" (Washington, D.C., Resources for the Future, 1983) p. 21.
[a]Excludes Alaska.

**Table 2–2. Per Capita Distribution of Public Recreation Land Holding Population Constant at 1980 Levels, 1960 and 1980**
(recreation acres per capita)

| Region | 1960 | 1980 | Change |
|---|---|---|---|
| New England | 0.15 | 0.16 | +0.01 |
| Mid Atlantic | 0.10 | 0.10 | 0.00 |
| East North Central | 0.14 | 0.17 | +0.03 |
| West North Central | 0.40 | 0.54 | +0.12 |
| South Atlantic | 0.18 | 0.31 | +0.03 |
| East South Central | 0.29 | 0.33 | +0.02 |
| West South Central | 0.23 | 0.27 | +0.04 |
| Mountain | 9.40 | 9.65 | +0.25 |
| Pacific[a] | 1.63 | 1.71 | +0.08 |
| United States | 0.87 | 0.92 | +0.05 |

*Note*: Includes only areas specifically designated for parks and recreation.
*Source*: Henry L. Diamond and coauthors, "Outdoor Recreation for America" (Washington, D.C., Resources for the Future, 1983) p. 22.
[a]Excludes Alaska.

breakdown of recreation land expressed in terms of the acres of recreational land per acre of all land (that is, the fraction of land devoted to recreation). With recreation availability measured in this way, the disparity between the highest and lowest regions is reduced from a factor of about one hundred to a factor of about ten. And while the Mountain and Pacific regions are still ranked at the top, the lowest-ranked regions are in the central part of the country—the West North and West South Central regions—with the East North and East South Central regions only slightly better off. With the data examined in this manner, improvement in the past two decades can no longer be neatly divided between East and West.

**Table 2–3. Public Recreation Land, 1960 and 1980**
(acres per acre of total land area)

| Region | 1960 | 1980 | Change |
|---|---|---|---|
| New England | 0.042 | 0.047 | +0.005 |
| Mid Atlantic | 0.057 | 0.056 | −0.001 |
| East North Central | 0.036 | 0.045 | +0.009 |
| West North Central | 0.021 | 0.028 | +0.007 |
| South Atlantic | 0.052 | 0.064 | +0.012 |
| East South Central | 0.036 | 0.042 | +0.005 |
| West South Central | 0.019 | 0.023 | +0.004 |
| Mountain | 0.914 | 0.919 | +0.005 |
| Pacific[a] | 0.243 | 0.251 | +0.008 |
| United States | 0.085 | 0.090 | +0.005 |

*Note*: Includes only areas specifically designated for parks and recreation.
*Source*: Computed using data from Henry L. Diamond and coauthors, "Outdoor Recreation for America" (Washington, D.C., Resources for the Future, 1983).
[a]Excludes Alaska.

Recall from chapter 1 that the one characteristic of an indicator of resource availability is that it must indicate the sacrifices that must be endured to enjoy a unit of the resource. Putting aside until later the question of what a "unit" means in the context of recreation, we may ask, what is likely to be a more accurate indicator of the sacrifice required, acres per capita or acres per acre? Acres per capita might be a rough indicator of the extent of crowdedness at recreation facilities, while acres per acre might measure the travel required to get to a recreation site. If sites are distributed randomly, the average distance to the nearest site decreases as this ratio increases. The problem with such measures as acres per acre or acres per capita is that they merely imply the costs faced by participants. The usefulness of ranking regions by criteria such as these is thus questionable, since the absolute number of acres per capita reveals little about the costs imposed on individual participants.

In addition, when we look beyond the aggregate figures to the quantity of particular kinds of facilities, we find that the West is not always better off than the East, even on a per capita basis. For example, table 2–4 shows the availability of golfing facilities, expressed in terms of the population per 18-hole course available to the public. Golf courses are most available in the North Central states; they are least available in the South Central states. The Pacific states, which score relatively high on aggregate acreage available, are well below average with respect to golf courses. Indeed, Alaska, which has by far the greatest amount of recreational land per capita, ranks forty-seventh among the states in golf course availability.

Both acreage per capita and acreage per acre also suffer because they give no information about how much the area under consideration is used. For this we need information on attendance, which can provide yet another perspective on recreation availability. Consider table 2–5, which gives regional attendance per acre at state parks and

Table 2–4. Availability of Public 18-Hole Golf Courses by Region, 1982

| Region | Population per 18-hole course open to the public |
|---|---|
| New England | 31,511 |
| Middle Atlantic | 45,626 |
| East North Central | 27,967 |
| West North Central | 30,823 |
| South Atlantic | 39,170 |
| East South Central | 62,091 |
| West South Central | 68,588 |
| Mountain | 31,986 |
| Pacific | 49,473 |

*Source*: Marion Clawson and Carlton Van Doren, eds., *Statistics on Outdoor Recreation* (Washington, D.C., Resources for the Future, 1984) p. 359.

**Table 2–5. National and State Park Attendance, 1981**
(annual visitation per acre)

| Region | National park units[a] | State parks |
|---|---|---|
| New England | 85.7 | 45.9 |
| Mid Atlantic | 109.3 | 255.6 |
| East North Central | 4.5 | 139.3 |
| West North Central | 13.9 | 72.3 |
| South Atlantic | 27.9 | 78.5 |
| East South Central | 24.0 | 220.0 |
| West South Central | 6.4 | 108.5 |
| Mountain | 3.3 | 39.1 |
| Pacific[b] | 5.9 | 116.5 |
| United States[b] | 9.6 | 104.0 |

Source: Marion Clawson and Carlton Van Doren, eds., *Statistics on Outdoor Recreation* (Washington, D.C., Resources for the Future, 1984) pp. 176, 179, and 242–243.
[a]Includes all units administered by the National Park Service (national parks, monuments, historical parks, and so forth).
[b]Excludes Alaska.

units administered by the National Parks Service (excluding Alaska). This table suggests that the differences among regions are less important than differences between state parks and national parks. In fact, state park attendance per acre is on average ten times that of national parks. (Total state park visitation is almost three times that of the national park units, on less than a quarter of the acreage.) But does this mean that state parks are too crowded? Or more popular? Or are they simply more accessible than the national parks? Or, finally, does the typical state park offer an essentially different recreational experience from that of the national parks?[1] In any case, comparisons based simply on visitation per acre do not easily support inferences regarding recreation resource availability.

## Effective Acres

Recently a more sophisticated measure of recreation supply was suggested by Marion Clawson in what he called "effective acres equivalent" (Clawson, 1984). This measure attempts to correct the raw acreage figures for the fact that some acres are used more than others.

[1]Certainly this last explanation would fit in with the Clawson-Knetsch (1966) classification of recreation areas. Clawson and Knetsch proposed classifying recreation areas into three groups: "resource-based" areas containing outstanding scenery or natural landmarks; "population-based" sites located near population centers but otherwise undistinguished; and "intermediate" areas containing elements of the other two. National and state parks were classified respectively as resource-based and intermediate areas.

He illustrates the concept with a hypothetical set of identical recreation sites, each of 200 acres, located at varying distances from a center of population. Visitation rates and travel costs to these sites are shown in table 2–6.

The important thing to note from the table is that visitation rates to the recreation sites decline with distance from the population center. In effect, the increased travel cost to the more distant sites acts as an entry fee to reduce participation.[2] Therefore, an acre at the more distant sites is less useful for recreation and should be discounted relative to an acre at the nearest site. Clawson suggests the discounting be based on the annual visitation rates per capita. Thus, the effective acreage equivalent (EAE) of the nearest site is the full 200 acres. The EAE of the next site is only 126 acres, or 44/70 of 200 acres. One rationale for this procedure is that, in terms of its actual use by recreation participants, 200 acres twenty miles away from the population center is "equivalent" to 126 acres ten miles away.

**Table 2–6. Number of Visits to Various Recreation Areas**

| Site | Distance to population center (in miles) | Annual number of visits per 1,000 population | Effective acre equivalent (EAE) |
|---|---|---|---|
| 1 | 10 | 70 | 200 |
| 2 | 20 | 44 | 126 |
| 3 | 30 | 28 | 80 |
| 4 | 40 | 16 | 46 |
| 5 | 50 | 7 | 20 |
| 6 | 60 | 0 | 0 |

*Source*: Marion Clawson, "Effective Acreage for Outdoor Recreation," *Resources*, no. 78 (Fall) p. 3.
*Note*: All sites are 200 acres large.

Clawson then applies the EAE concept to the recreation acreage in state and national parks, national forests, and county and city parks, and concludes that the effective acreage of local parks actually exceeds that of national parks, even though the acreage of local parks is only 6 percent of that of national parks. This calculation is a quantitative statement of the fact that the vast bulk of recreation activity takes place on only a small part of recreational land. Because it takes accessibility into account, the EAE is a significant improvement over raw acreage as a measure of total resource availability.

The EAE can also take into account the effects of congestion on consumer satisfaction—at least to the extent that users stay away on that account—since it weighs acreage by per capita attendance. Suppose regional population doubles and attendance at a given site re-

[2]This insight is also the basis of the immensely influential "travel cost" method, developed by Clawson (1959) as a means of valuing recreation resources.

mains the same. Use per capita would then be reduced by half, which would show up as a decline in effective acreage at that site. The EAE also has the virtue of being conceptually straightforward and easy to calculate.

The EAE also has some flaws, however, that may make it uninformative as an indicator of recreation supply. For one thing, it is an *ad hoc* adjustment to differences in site accessibility, not grounded in any theory. In particular, it offers no insight into how acres or any other input contribute to recreation supply. Indeed, one could, with equal justification, weight sites, rather than acres, by per capita visitation. In addition, the weighting process disconnects the final measure from any physically or economically meaningful concept, because the determination of the baseline is arbitrary. The EAE measure is an index only. Furthermore, although the EAE usefully summarizes in one index several aspects of recreation resource availability, in a sense it summarizes too much. Consider the above example, where population doubles but attendance remains the same. Is this a decline in resource availability induced by congestion, or simply a shift in the demand curve?[3] The difference is important for policy, but is not suggested by the EAE calculation.

## Effective Price

A meaningful quantitative assessment of the nation's recreation resources is difficult to establish. Current measures are based almost entirely on physical measures, mostly of land area. These measures are simple in concept and the data necessary to calculate them are abundant, if not always conveniently assembled. However, they reveal as much about the supply of recreation as, say, the number of paintings reveals about the supply of art.

Recreation scarcity measures based on acreage assume implicitly that more acres are better. While this is almost certainly true, it fails to consider exactly how the addition of recreation acres increases recreation opportunities. A moment's thought suggests several reasons why one might want to expand the stock of recreation resources in a region. First, there are many different kinds of recreation activities. If there are no tennis courts in a neighborhood, for example,

[3]Lest this be thought to be an artificial example, consider that since 1978, attendance at the urban units of the National Park System has increased 65 percent, compared to a 2 percent increase in attendance at nonurban parks (interview by author with Kenneth Hornbeck, National Park Service Data Center, Denver, Colorado, April 1985). Evidently, the use of the urban parks has increased relative to the nonurban parks. The reason for the shift is unclear.

there may be unmet recreational demands even though there are ample recreation facilities of other types. Adding a recreation site also reduces the distance traveled for some participants and may reduce crowding, as some choose the new site over existing ones.

A meaningful scarcity measure must consider these elements explicitly, rather than indirectly or not at all, as in the case of physical measures. In the remainder of this study, it will be argued that travel and congestion costs can be converted, at least in principle, to an effective price that provides a measure of recreation resource scarcity.

# three

# PRODUCTION OF RECREATION OUTPUTS AT A SINGLE SITE

In this chapter, an attempt will be made to develop a production theory for recreation. Specifically, a cost function for a recreation site will be defined and used to examine the ability of a recreation site to produce outputs under several different site management policies. But just what are the "outputs" or services produced by a recreation site, and how are they measured? Definition of output is essential to the definition of production or cost functions, and this definition depends on how individuals derive satisfaction from recreation.

## Defining Outputs

In general, individuals derive satisfaction from recreation sites in two ways. First, they can receive "nonuser benefits" from a park without ever visiting it, valuing the fact that the park exists should they ever decide to visit it (option value), or valuing the park simply because it exists (existence value). Certainly, the "crown jewels" of the National Park System (such as the Grand Canyon and Yosemite) have this sort of nonuser value. In this context, a park is a pure Samuelsonian public good, for the park itself directly enters the utility function of each person. As such, it possesses two characteristics not found in a typical private good: nonexcludability (it is impossible to prevent anyone from enjoying the park's existence) and nonrivalness (one individual's enjoyment of the park's existence does not interfere with another's).

Nonuser benefits will not be considered further. Rather, the discussion will center on the second category of park benefits, the experiences enjoyed by site visitors. What we call "experiences" are commonly measured in units of participation such as visits, visitor days, or participation days. In the production of recreation experiences, site acreage is only one input among many, which is why various measures of acreage are inadequate as indicators of recreation supply. At the site, the operator must provide additional inputs that vary greatly, depending on the type of recreational activity and the quality of the experience. Downhill skiing, for example, requires a hill covered with snow, lifts, and lodge facilities, at a minimum. In contrast, the operator of a hiking trail must provide relatively little besides land to produce opportunities for hiking.

The participant also provides inputs, such as the leisure time allotted to the experience, travel, plus specialized recreation equipment (skis, powerboats, and so forth). To be sure, time and travel are required for the purchase or consumption of almost any good or service, but for most goods they account for a relatively small part of the total cost. In contrast, time and travel are important inputs to the production of recreation experiences.

Because participants as well as operators of recreation areas provide inputs into the production of recreation experiences, a household production function is suggested. The following simple model shows how the operator's inputs affect the household's allocation decisions, using a household production function approach in the style of Becker (1965).

Suppose household utility $U = U(R, X; Q)$, where $R$ is the number of recreation experiences enjoyed (per unit time), $X$ represents all other goods, and $Q$ is an index of the quality of the experience, exogenously determined by the operator of the recreation area and initially set at $Q = Q_0$. The household maximizes utility subject to constraints on income and available time, so that the problem is to maximize

$$U = U(R, X; Q_0) \tag{3-1}$$

subject to

$$p_R R + p_x x = Y = wW \tag{3-2}$$

and

$$W + \tau R = T \tag{3-3}$$

In equation (3–2), $p_R$ is the money price of a recreation experience, including the (amortized) equipment cost, transportation costs, and entry fees; $p_x$ is the price of $x$; $Y$ is the income; $w$ is the wage rate; and $W$ is the total number of hours worked. In equation (3–3) $\tau$ is the time devoted to each recreation experience and $T$ the total time available. Combining equations (3–2) and (3–3) to form a single constraint, the Lagrangian is

$$L(z, x, \lambda; Q) = U(R, X; Q) + \lambda[wT - p_x x - (p_R + wt)R] \qquad (3\text{–}4)$$

The first-order conditions are

$$\frac{U_x}{\lambda} = p_x; \quad \frac{U_R}{\lambda} = p_R + w\tau \qquad (3\text{–}5)$$

Passing to parameter space in the usual fashion, we define the demand function for recreation $R(p, Q)$ and the indirect utility function $V(p, Q)$, where $p$ is the money price of the recreation experience, initially $p_R$, and $Q$ its quality, initially $Q_0$ (other parameters are assumed unchanged). Because quality differences are difficult to deal with analytically, it is customary to convert such differences into an "equivalent" change in the price by finding the change in price that has the same effect on utility as the quality change. Thus, if quality decreases from $Q_0$ to $Q_1$, there corresponds a price change $\Delta^V(Q_1)$ such that the decrement in utility when quality decreases from $Q_0$ to $Q_1$, with price at the initial level of $p_R$, is the same as the decrement when the price increases from $p_R$ to $P_R + \Delta^V(Q_1)$, with quality at $Q_0$. In terms of the indirect utility function $V(p, Q)$, this can be expressed as follows

$$V(p_R + \Delta^V(Q_1), Q_0) = V(p_R, Q_1) \qquad (3\text{–}6)$$

The difference $\Delta^V(Q_1)$ is a monetized expression for measuring the difference in quality. This is essentially the method used, for example, by Freeman and Haveman (1977) to value quality changes due to congestion.

But this is not the only way to monetize differences in quality and, in fact, is not the most useful for our purposes. The problem is that the recreation price change that keeps utility constant for a change in quality many not result in the same recreation frequency. In other words, if $R(p, Q)$ is the demand function for recreation experiences, then $R(p_R + \Delta^V(Q_1), Q_0)$ is not necessarily equal to $R(p_R, Q_1)$. Because our interest is in the allocative role of quality, it is desirable to have a measure that leaves the quality demanded unaffected. Accordingly, we define the measure of quality so that quantity is unchanged. If quality decreases from $Q_0$ to $Q_1$, then let $\Delta^R(Q)$ be defined by the equation

$$R(p_R + \Delta^R(Q_1), Q_0) = R(p_R, Q_1) \tag{3-7}$$

The quality difference is measured by $\Delta^R(Q_1)$.

Ordinarily, the demand of a good is a decreasing function of its price, but this is an empirical outcome and not a logical consequence of the household utility maximization model. Similarly, it is not logically necessary that demand be an increasing function of quality.[1] Nonetheless, throughout this paper it is assumed that recreation frequency is an increasing function of the quality of the experience. Thus $\Delta^R(Q_1) > 0$ in equation (3–7) and the cost to the household of a quality decrement is positive.

Now the components are available to define the effective price of a recreation experience. Recall from the first-order conditions given in equation (3–5) that the price of an ordinary good equals the ratio of marginal utility to the Lagrange multiplier. A similar ratio for recreation gives an expression that has a price-like allocative function, with $Q_0$ as a reference quality level. The effective price of a recreation experience, given the parameters $p_R$ and $Q_1$, is

$$p = p_R + w\tau + \Delta^R(Q_1) \tag{3-8}$$

The components of this effective price are

  (a) the money price ($p_R$)
  (b) the monetized value of the time devoted to the experience ($w\tau$), including travel time and time on the site
  (c) the (relative) quality of the experience, $\Delta^R(Q_1)$

Equation (3–8) decomposes the effective price according to the "currency" units in which payment is made: money, time, and quality. However, there is a different decomposition of the price that is more useful for our purposes: the entry fee, travel cost, congestion cost, and intrinsic site quality. Travel cost includes both a money element and a time element. Congestion cost, as noted above, may include a time element or a quality element, or both. Each of the three terms in equation (3–7) can be affected by actions of the operator. The money price is affected by the entry fee. The time component can be affected by the proximity of recreation sites and the facilities at those sites for handling crowds.

[1] A good $x$ for which $x_Q < 0$ may be more common than Giffen goods (for which $X_p > 0$). An example might be automobile tires. Tire quality is mainly exhibited in durability, and longer-lasting tires need to be replaced less often.

At a congested facility, for instance, participants may have to endure more time waiting in line and have correspondingly less time to devote to the activities that give pleasure. Indeed, the quality of the experience may largely be determined by the amount of time granted the latter (downhill skiing is an example). Thus, in addition to lowering the entry fee, numerous other actions, such as expanding the capacity of the facility to reduce congestion, building new sites closer to population centers, or increasing the quality of the experience, can be taken by the operator to lower the effective price.

For the balance of this chapter, our focus will be on the ability of a recreation area to produce recreation experiences, assuming no travel costs. Incorporation of travel costs requires all recreation areas to be considered simultaneously, and will be introduced in chapter 4. Introduction of intrinsic site quality also makes more sense if done in the multiple-site context. For the most part, it is assumed here that all recreation areas offer the same quality level (except for congestion); this assumption will be relaxed at the end of chapter 4.

For the remainder of this chapter, the focus will be on the costs incurred on the site (by both the recreation participants and the site operator). For the participants, a major element of the cost incurred on site is the cost of congestion. The nature of congestion cost—how congestion affects utility, how congestion is affected by site inputs, whether congestion increases cost or decreases the quality of an experience and whether it matters—is an important issue that will be considered in some detail in chapter 5. For now, congestion cost is treated as if it were given and similar to any other cost.

## Defining a Cost Function

The user services provided by recreation sites often possess the public good attributes of nonexcludability and nonrivalness, even though the final service produced—the recreation experience—is private in the sense that an individual's recreation experience enters only his own utility function. As Head (1962) has observed, neither nonexcludability nor nonrivalness implies the other, and neither is an inherent property of a good, but rather depends on transaction costs or the level of use. Although few recreation resources are inherently nonexcludable, for some, such as city parks, exclusion would be difficult and expensive. For many others, nonexclusion is practiced as a matter of policy. Thus, for historical and political reasons the recreation units of the National Park Service and Forest Service generally do not ration access by price. The entry fees charged, if any, are much less than required to reduce congestion. (At some facilities exclusion on the basis of a lottery, reservation system, or first come-first served must be practiced, but the philosophy of the Park Service is that access should not be determined by ability to pay.) Likewise, recreation resources are often nonrival, because one individual's use of the site does not interfere with its use by others as long as the use level is sufficiently low. But as visitation increases, the same participants will eventually feel that their satisfaction is diminished by the presence of others, at which point the service flow is no longer nonrival.

Those properties of nonexcludability and nonrivalness make it difficult to associate an output with a particular set of inputs, which is, of course, the definition of a production function. Indeed, nonexcludability suggests that the producer does not even retain control over the output (the number of recreation experiences). Nonrivalness means, furthermore, that the same recreation resources can support many different levels of use, depending on how many visitors show up. And even at a congested facility, the same inputs can provide many different levels of output, though not of the same satisfaction to the user. Because of this indeterminacy between the level of inputs and the level of output, no production function is defined. Rather, we shall define a cost function that has as arguments the site inputs plus the level of output—the number of visitors or recreation experiences.

Let the inputs at a recreation site be denoted by the vector $A = (A_1, A_2, ..., A_n)$. These include land, buildings, equipment, such as ski lifts, and so forth. These inputs are assumed to be under the control of an "operator." At a private recreation site, the operator is the owner or manager. At a public site, the operator is the public agency in charge.

Investment by the operator can enhance recreation opportunities in two ways, as shown in figure 3–1. First, the cost of congestion can be reduced, by adding land or by building new roads or picnic tables to handle a larger crowds. This is represented by the shift from $Q(N, A_0)$ to $Q(N, A_1)$. Alternatively, the congestion-free quality can be improved, shifting the willingness-to-pay curve from $Q(N, A_0)$ to $Q(N, A_2)$.[2] Note that as drawn, the congestion-free quality improvement has more effect on willingness to pay for $N < N_0$, whereas the congestion-reducing improvement has more effect for $N > N_0$. In making investment decisions, the operator needs to know the tastes of his clientele and the level of use.

The cost of these inputs is the cost to the operator of producing recreation experiences. As noted above, the recreation participants also provide inputs into the production of a recreation experience,

---

[2]This may not be possible for scenic areas, where it is impossible to improve upon nature.

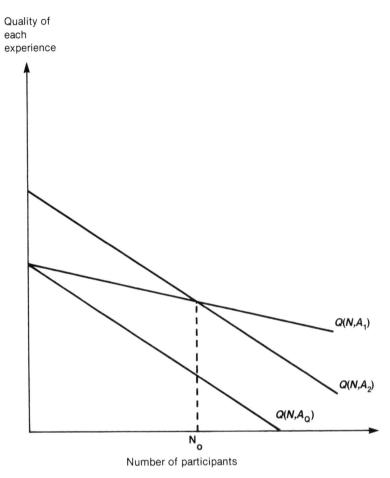

Quality of
each
experience

$Q(N,A_1)$

$Q(N,A_2)$

$Q(N,A_Q)$

$N_o$

Number of participants

**Figure 3–1.** Investment to improve quality in recreation

and these costs, plus the cost of congestion, must be included in the cost of production of recreation experiences.

Thus, the social cost of producing $N$ recreation experiences, for $N$ households, is shared between the owner or operator of the site and the households themselves. Let us assume that all participants have the same taste (or distaste) for congestion, and all have the same personal inputs. Then the social cost of producing $N$ recreational experiences simultaneously is a function of the operator's inputs, the vector $A$, and the number of participants $N$

$$
\begin{aligned}
T(N,A) = \ & K_0(N,A) && \text{(operator's cost)} \\
& + NK_p && \text{(personal inputs)} \\
& + NC(N,A) && \text{(congestion)}
\end{aligned}
$$

Several points should be noted about this cost function. First, travel cost is one of the "personal inputs" and is assumed to be the same for each individual. The interesting complications that arise when travel costs differ across individuals will be discussed in chapter 4. All other personal inputs will be ignored henceforth because they have no relevance for public policy. Second, the social costs are shared between the operator and the participants, as mentioned above. Third, the cost to the operator, $K_0$, is certainly a function of the amount of the input $A$ employed, and may depend on the number of participants $N$. The dependence on $N$ is usually indirect; as when the operator adjusts the inputs $A$ to the number of participants (for example, extra lifeguards at a pool on weekends). The operator costs may also depend on $N$ directly, as when excessive use causes facility damage. Fourth, the operator's inputs influence the congestion cost function. This is an obvious point and one that has been mentioned before, for congestion costs associated with a given participation rate can clearly be reduced by adding land or capital inputs, such as roads or campsites. But although this is well understood in a qualitative sense, very little is known about the *quantitative* effects of site inputs on congestion costs. Finally, the arguments of this cost function include both the inputs $A$ and the output $N$, a reflection of the fact that there is no unique relationship between inputs and outputs.

One other unusual feature of this cost function should be noted. It is not immediately obvious who controls the arguments, or at least the argument $N$ (the operator's control over the inputs $A$ seems clear enough). In some cases it is the operator who exercises ultimate control over the number of visitors $N$, albeit indirectly and sometimes imperfectly, through the admission policy. Admission to a facility may be regulated in two ways: directly, by a "quantity instrument"— a limit on the number admitted—and indirectly, by a "price instrument"—an entry fee (Weitzman, 1974b).

Often, however, the operator cannot, or will not, control the number of visitors. At such "open access" facilities, the level of output is determined not by the operator, but by the sum of the independent decisions of a larger number of recreation participants. Attendance is then rationed by the other costs facing the participants: the cost of travel, the scarcity of leisure time, the distaste for congestion. Open access is by far the most common policy for rationing public recreation facilities in the United States, not only for city parks and the like, where controlling access is difficult and expensive, but also for the national parks and monuments, where control of access is feasible.

When attendance is rationed by price or open access, the final decision on whether to attend is made by the participant. That is, the operator does not bar entry to the site the way he does with a quantity-based instrument. Instead, the operator's influence over output is indirect, through pricing policy or provision of site inputs $A_1, \ldots A_n$, and the effect of such inputs on the price and quality of the experience. This is quite different from conventional production theory, where the producer can select a target output independent of the demand curve for the product. In other words, the supply of recreation experiences is inextricably connected to the demand for those experiences; hence a conventional supply curve cannot be defined. Analysis of the relationship between recreation inputs and outputs thus requires two elements in addition to the cost function: the aggregate demand curve and the method, if any, by which access is controlled.

The demand curve $D(p)$ is the sum of the individual demand curves for recreation of each member of the population in the absence of congestion. To define it, we begin with demand, defined, as in the preceding section, as a function of the price and quality of the recreation experience. The quality of the experience is determined entirely by congestion, which in turn is assumed to depend on the number of participants at the site. Thus, for a given member of the population, the number of recreation visits is assumed to be a function of the entry fee $p$ and the number of other participants $N$ found at the site, $R(p,N)$. In this demand curve, the quantity is measured by the frequency of participation per unit time. For example, if at price $p$ and usage level $N$, the individual's demand for service is ten visits per month, on a given day the expected number of visits is one-third.

Let us assume $R_1 = \partial R(p,N)/\partial p < 0$ (the standard assumption of downward-sloping demand) and $R_2 = \partial R(p,N)/\partial N < 0$, so that as the facility becomes more crowded its attractiveness diminishes. Suppose further that the function $R(p,N)$ is such that there is a function $C(N)$ that satisfies the following condition

$$R(p,N) = R(p + C(N), 0) \qquad (3\text{--}9)$$

for all $p$ and $N$. The function $C(N)$ is called the congestion cost function, and has the following properties:

(a) $C'(N) = \dfrac{R_2(p,N)}{R_1(p + C(N),0)} \geq 0;$

(b) $C(0) = 0;$ and

(c) $C(N) \geq 0$ for all $N$

With congestion cost defined in this way, the constant-congestion demand curves differ by a constant, so that the congestion cost $C(N)$ is independent of the fee $p$. This assumption allows the congestion cost to be treated as any other cost to the participant. Indeed, when

$N$ other participants are present at the site and the entry fee is $p$, the individual behaves as though no others are present and the fee is $p + C(N)$.

Usually, the site operator can increase site "capacity" by adding land, or perhaps by building new roads or picnic tables to handle larger crowds. When site inputs are added in this fashion, the congestion cost experienced by an individual with $N$ other participants present should decline. Denote these site inputs by a single variable $A$. Because the congestion cost depends on $A$ as well as $N$, the congestion cost function can now be written as $C(N,A)$. In addition to $C_N > 0$, we assume $C_A < 0$ and $C(N,A) \rightarrow 0$ as $A \rightarrow \infty$.

If all $U$ households in the population have the same demand curve and congestion cost function, the aggregate congestion-free demand curve is

$$D(p) = UR(p,0)$$

$D(p)$ is thus the expected number of users in a day (or any other interval) when the entry fee is $p$, assuming no congestion. Since by definition the congestion experienced acts just like an increase in price, the demand curve for a site of capacity $A$, with entry fee $p$ and attendance $N$ is

$$D(p + C(N,A)) \tag{3-10}$$

That is, $D(p + C(N,A))$ is the expected number of recreation experiences that would be demanded at a site if the entry fee were $p$ and the congestion cost were $C(N,A)$. For fixed $N$ and $A$ this defines a constant-congestion aggregate demand curve. Note that only one point on this curve can be observed, namely $D(p + C(N_0,A)) = N_0$.

The social cost of supplying $N$ recreational experiences is the sum of the costs to the participants and the operator's cost of providing and maintaining the facility (entry fees, if any, net out). Neglecting the costs of travel and personal equipment, each participant's cost is the cost of congestion, or $C(N,A)$. Suppose the operator's costs are a fixed amount per unit time, irrespective of the number of users at the site. Then the cost of $N$ experiences within a given time period is

$$T(N,A) = NC(N,A) + K_0(A) \tag{3-11}$$

## Conditions Defining Size and Level of Use

With the cost function $T(N,A)$ given by equation (3–11) and the aggregate demand $D$ given by equation (3–10), only the operator's management regime—that is, whether attendance is regulated by the operator and if so, whether by a price or quantity instrument—is needed to determine the quantity of recreation experiences demanded and the size of the facility.

Let us contrast three different admission policies, namely, the economically efficient policy, the revenue maximizing policy, and an open access policy, in which attendance is unrestricted by the operator.

In the short run, the level of inputs $A$ at the site is fixed, and comparison of attendance under the three regimes leads to the well-known result that the free access attendance is greater than the economically efficient attendance, which in turn is greater than the revenue maximizing attendance. This result is illustrated in figure 3–2. If price is zero and access is free, individuals come as long as their willingness to pay exceeds their congestion cost, $C(N,A)$. That is, one's participation decisions reflect only congestion imposed on the individual, not the incremental congestion costs imposed by the individual on others. The open-access equilibrium $\overline{N}$ occurs where the demand curve and the congestion curve $C(N,A)$ cross (in panel A of figure 3–2).

Now suppose the operator aims for the economically efficient result by means of an entry fee, setting the fee to maximize the benefit function $B(N)$, where

$$B(N) = \int_0^N D^{-1}(q)\,dq - T(N,A) \tag{3-12}$$

In equation (3–12), $D^{-1}$ denotes the inverse demand function and the integral represents the total consumer surplus in the absence of congestion. The optimum $N$, $N^*$, occurs where the marginal aggregate congestion cost curve, determined by differentiating $T(N,A)$ in equation (3–11) with respect to $N$, crosses the demand curve, as shown in panel B of figure 3–2. Note that the congestion suffered by each participant at the optimum, $C(N^*,A)$, is less than the marginal willingness to pay, which is shown as $W^*$. To achieve the optimum result, access must be limited by a restriction on $N$ or a fee $p^* = N^*C(N^*,A)$ must be charged. It should be noted that the use of a quantity instrument rather than a fee for controlling access may prevent the efficient result from being achieved. For example, if late arrivals must queue up after the allowable participation rate is achieved, then the difference between $W^*$ and $C(N^*,A)$ is a deadweight loss. If admission is random then the quantity instrument can achieve the optimum result, but only if all individuals have identical demand curves. Both Mumy and Hanke (1975) and Porter (1977) discuss these nonprice allocation mechanisms.

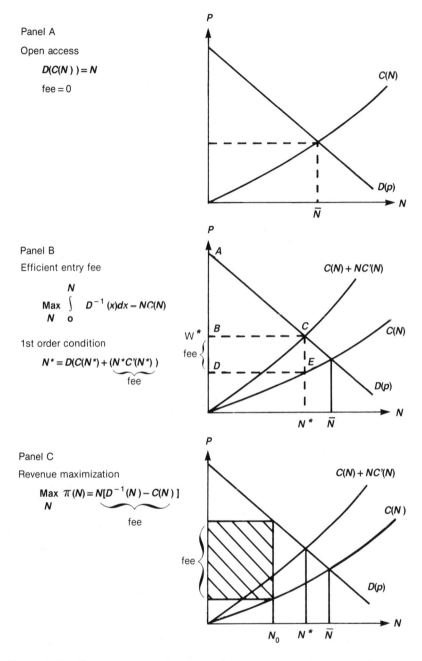

**Figure 3–2.** Open-access, optimum, and revenue-maximizing attendance at a recreation site, with site inputs fixed

Because the average congestion cost curve is exceeded by the marginal curve, $N^* < \bar{N}$. This is a standard result in urban economics, where congestion issues often arise (see, for example, Mills, 1972, p. 201).[3] Note also that the gain to the participants themselves is maximized at the open access equilibrium. This is another standard result (see Weitzman, 1974a; or Baumol and Oates, 1975, p. 194) and follows from comparison of the areas $ABC$ and $ADE$ in panel B of figure 3–2.[4] This conclusion must be tempered by consideration of where the collected fees end up. If, for example, everyone in the society is a recreation participant, then the fees are returned to the participants, and this conclusion is not valid. On the other hand, if there are a substantial number of nonrecreators who share in the collected fees, it is less likely that the participants will be better off with optimum fees.

When the operator acts as a monopolist and imposes a fee designed to maximize revenue, the optimum $N$, $N_0$, is less than the efficient level $N^*$. $N_0$ is found by maximizing the revenue function

$$\Pi(N) = N[D^{-1}(N) - C(N,A)]$$

$N_0 < N^*$ follows from the fact that the total benefit function $B(N)$ has an inverted U-shape and $B'(N_0) > 0$, so that $N_0$ lies to the left of the efficient point $N^*$. This is illustrated in panel C of figure 3–2.

## Long-run Participation and Input Levels

In the long run, both the participation rate and the quantity of acreage depend on the management strategy. Let us assume the operator wishes to maximize economic efficiency and again compare two cases, one in which the operator optimizes over both $A$ and $N$ (the global optimum) and a second best alternative in which the operator selects the size of the facility only, with $N$ determined by the open-access equilibrium. It is also assumed that the operator's cost function is linear, $K_0(A) = wA$, so that the cost of adding site inputs is constant. For convenience, let us assume only one input, land, with $w$ therefore the implicit rental price of land. Consider first the case in which the

[3] As Freeman and Haveman (1977) have noted, this conclusion depends on the assumption that everyone has the same taste for congestion. Also, if there are many sites it does not necessarily follow that the open-access participation is greater than the optimum participation *at each site* (De Meza and Gould, 1985).

[4] Again, this conclusion does not follow when individual tastes are heterogeneous. Users with a high aversion to congestion would prefer a pricing scheme that dissuades all but a few users to an open access scheme.

operator can control the visitation rate. The benefit function is given by the difference between consumer surplus and the aggregate cost function $T(N,A)$.

$$B(N,A) = \int_0^N D^{-1}(q)dq - T(N,A) \qquad (3\text{--}13)$$

where $D^{-1}$ is the inverse demand function. The first order conditions for solving equation (3–13) are

$$B_A = -w - NC_A(N,A) = 0 \qquad (3\text{--}14a)$$
$$B_N = D^{-1}(N) - C(N,A) - NC_N(N,A) = 0 \qquad (3\text{--}14b)$$

The first of these conditions states that the reduction in congestion cost due to adding an acre of land multiplied by the number of users must equal the rental price of land. The second condition states simply that price must equal the increase in congestion cost due to the last participant, which is the sum of one's own congestion cost ($C(N,A)$) plus the increase in other users' congestion cost ($NC_N$).

If the operator cannot or will not restrict visitation, this optimum cannot be achieved. However, the operator can aim for a long-run second best by maximizing over acreage only. When attendance is not restricted by the operator, it rises to the level at which the average congestion cost and demand curves intersect, as shown in panel A of figure 3–2. Thus, for any area $A$, there is a participation rate $N$ where these curves intersect. In a sense, attendance is being rationed by congestion in this case. This rate is found by solving the pair of equations

$$D(p) = N$$
$$p = C(N,A)$$

for the unknowns $p$ and $N$. More compactly, $p$ can be eliminated from this system to yield a single equation

$$D(C(N,A)) = N \qquad (3\text{--}15)$$

which defines $N$ implicitly as a function of $A$ that will be called $\overline{N}(A)$. Note that

$$D'(C(\overline{N},A))(C_A + C_N\overline{N}'(A)) = \overline{N}'(A)$$

or

$$\overline{N}'(A) = \frac{D'(C(N,\overline{A}))C_A}{1 - D'(C(\overline{N},A))C_N}$$

Since $D' < 0$, $C_A < 0$, and $C_N > 0$, $\overline{N}'(A) > 0$. As acreage increases at a free access site, equilibrium attendance also increases. Intuitively, we also expect $\overline{N}'(A) \to 0$ as $A \to \infty$, for as recreation area expands, demand is eventually sated. If $D'$ is bounded this certainly is true because $\overline{N}'(A) < D'(C(\overline{N},A))C_A \to 0$ as $A \to \infty$ (because $C_A \to 0$). Now the operator optimizes

$$\overline{B}(A) = B(A,\overline{N}(A)) \tag{3–16}$$

$$= \int_0^{\overline{N}(A)} D^{-1}(q)dq - \overline{N}(A)C(\overline{N}(A),A) - wA$$

The first-order condition is

$$\overline{B}'(A) = D^{-1}(\overline{N}(A))\overline{N}'(A) - \overline{N}'(A)C(\overline{N}(A),A)$$
$$- \overline{N}(A)\frac{d}{dA}C(\overline{N}(A),A) - w = 0$$

Since $N = D(C(N,A))$, $D^{-1}(N) = C(N,A)$, so that the first two terms of this expression are equal, leaving

$$\overline{B}'(A) = -\overline{N}(A)\frac{d}{dA}C(\overline{N}(A),A) - w = 0 \tag{3–17}$$

or

$$-\overline{N}(A)[C_A + C_N\overline{N}'(A)] - w = 0$$

In words, the optimum acreage occurs where the product of the attendance and the marginal reduction in congestion cost due to the additional area equals the cost of using land. This resembles one of the first-order conditions for optimizing over both $N$ and $A$, equation (3–14a), but in this case, the total derivative replaces the partial derivative. For a given $A$ the total derivative, $d/dA\ C(\overline{N}(A),A)$, is smaller in absolute value than the partial derivative, because additional acreage lowering congestion cost induces greater participation, which increases congestion cost.

Solution of equation (3–17) yields $\overline{A}$, the optimum recreation acreage when participation is restricted only by mutual congestion. The corresponding $\overline{N} = \overline{N}(\overline{A})$ is the equilibrium participation at this acreage. It is interesting to compare this solution to the "first best" solution $(N^*,A^*)$, found by solving equations (3–14a) and (3–14b). We know that for a fixed $A$ the equilibrium unrestricted $N$ exceeds the optimum $N$. Thus, it seems reasonable to expect that the second best $N$, $\overline{N}$, would exceed the optimum optimorium $N^*$, because access is not restricted. One might further expect the second best area $\overline{A}$ to exceed the optimum $A^*$, in recognition of the greater participation. However, neither of these two statements turns out to be true in general. Wilson (1983) has shown that the access area $\overline{A}$ will be less than the optimum $A^*$ if the demand is sufficiently elastic with respect to the price. That result does not guarantee that $\overline{N}$ will be less than $N^*$, since the open-access use rate always exceeds the optimum by $N^*$, is shown by the following example.

## Example

Suppose that the recreation demand function is given by

$$D(p) = \begin{array}{l} k/p^\alpha(\alpha > 0,\ 0 < p \le \overline{p}), \\ 0 \text{ otherwise} \end{array} \tag{3-18}$$

(The demand curve is truncated at $\overline{p}$ to ensure integrability.)
Suppose also that the congestion cost function is given by

$$C(N,A) = \frac{\theta N}{A} \tag{3-19}$$

With these functions the first-order conditions (3–14a) and (3–14b) for the overall optimum become

$$B_A = -w + \theta N^2/A^2 = 0 \tag{3-20a}$$

$$B_N = \frac{k^{1/\alpha}}{k^{1/\alpha}} - \frac{2\,\theta N}{A} = 0 \tag{3-20b}$$

After some arithmetic, we have

$$A^* = \frac{k\,\theta^{(1-\alpha)/2}}{2^\alpha w^{(\alpha+1)/2}} ;\ N^* = \frac{k}{2^\alpha\,\theta^{\alpha/2}w^{\alpha/2}} \tag{3-21}$$

Now suppose attendance is not restricted by the operator. Attendance as a function of acreage satisfies equation (3–15), which in this case becomes

$$N = D(C(N,A)) = \frac{kA^\alpha}{\theta^\alpha N^\alpha}$$

from which

$$\overline{N}(A) = \frac{k^{\frac{1}{1+\alpha}} A^{\frac{\alpha}{1+\alpha}}}{\theta^{\frac{\alpha}{1+\alpha}}} \tag{3–22}$$

and

$$\overline{C}(A) = C(\overline{N},A) = k^{\frac{1}{1+\alpha}} \theta^{\frac{1}{1+\alpha}} A^{\frac{-1}{1+\alpha}} \tag{3–23}$$

Applying equation (3–17) we have, after more arithmetic,

$$\overline{A} = \frac{\theta^{(1-\alpha)/2} k}{(1+\alpha)^{(1+\alpha)/2} w^{(1+\alpha)/2}} \tag{3–24}$$

and

$$\overline{N} = \overline{N}(\overline{A}) = \frac{k}{[(1+\alpha) w \theta]^{\alpha/2}} \tag{3–25}$$

Consider now the ratio

$$\frac{\overline{A}}{A^*} = \frac{2^\alpha}{(1+\alpha)^{(1+\alpha)/2}} \quad \begin{matrix} < 1 \text{ if } \alpha > 1 \\ = 1 \text{ if } \alpha = 1 \\ > 1 \text{ if } \alpha < 1 \end{matrix} \tag{3–26}$$

When $\alpha > 1$, $\overline{A} < A^*$, that is, optimizing with attendance unrestricted leads to a smaller area devoted to recreation than is globally optimal. Inasmuch as $\alpha$ in this model represents price elasticity, this is a special case of Wilson's (1983) result. In addition,

$$\frac{\overline{N}}{N^*} = \frac{2^\alpha}{(1+\alpha)^{\alpha/2}} \quad \begin{matrix} < 1 \text{ if } \alpha > 3 \\ = 1 \text{ if } \alpha = 3 \\ > 1 \text{ if } \alpha < 3 \end{matrix} \tag{3–27}$$

Again, $\overline{N}$ can be less than $N^*$ for suitably large values of $\alpha$.

In this example, the outcomes of interest depend only on the demand function, and, in particular, the parameter $\alpha$ (the demand elasticity). As the demand function becomes more and more elastic, $\overline{A}$ and $\overline{N}$ fall relative to their respective efficient levels, until the ratios

$\overline{A}/A^*$ and $\overline{N}/N^*$ drop below 1 when $\alpha$ exceeds 1 and 3, respectively. Note, however, that $\alpha \to 0+$, the ratios $\overline{A}/A^*$ and $\overline{N}/N^*$ both approach unity. Thus, for highly inelastic demand, it does not matter much which admission policy is used. Only for interim values of $\alpha$ can $\overline{A}$ exceed $A^*$ and $\overline{N}$ exceed $N^*$. Indeed, $\overline{A}/A^*$ takes its maximum value of 1.04 at about $\alpha = 0.5$, so that $\overline{A}$ never exceeds $A^*$ by much.

Above it was noted that when facility size is fixed, the benefit to each participant is maximized by an open-access policy. The present example can be used to show that this conclusion, too, does not generally hold in the long run. The participant's benefit is directly determined by the effective price: the congestion cost plus the entry fee, which is of course zero for the open-access solution and $N^*C_N(N^*,A) = (\theta w)^{1/2}$ for the efficient solution. Now consider the ratio of effective prices

$$\frac{C(N^*,A^*) + N^*C_N(N^*,A^*)}{C(\overline{N},\overline{A})} = \frac{2}{(1 + \alpha)^{1/2}} < 1 \qquad (3\text{--}28)$$

whenever $\alpha > 3$. When demand is extremely elastic, even the participants are better off with an efficient solution rather than the open-access solution.

The relationship between optimizing over both $A$ and $N$ and optimizing over $A$ only is illustrated in figure 3–3. This figure shows, in the $A - N$ plane, the isoquants of the benefit function $B(N,A)$ given in equation (3–11). As shown on the figure, $(N^*,a^*)$ is the global optimum, the choice of acreage and attendance that maximizes net benefits. The curve marked $B(N,A) = B_1$ is the set of points $(N,A)$ that yield net benefits of $B_1$ and similarly for $B_2$ and $B_3$. Because $(N^*,a^*)$ is the point of maximum benefits, $B(N^*,A^*) > B_3 > B_2 > B_1$.

Also in figure 3–3 is the curve $\overline{N}(A)$, the equilibrium attendance for area A when access is not limited. Because for a given $A$ the equilibrium $N$ exceeds the optimum $N$, we have $N^* < \overline{N}(A^*)$, so that the point $(N^*,A^*)$ must lie below the curve $\overline{N}(A)$. The second best area $\overline{A}$, the optimum area when access is unrestricted, is found by moving along the curve $\overline{N}(A)$ until one reaches the highest possible isoquant of the function $B(N,A)$, or $B_2$ in figure 3–3. $\overline{N}(A)$ is the tangent to the isoquant at this point.

As illustrated in the example above, the location of the point of tangency depends on the demand and congestion cost functions. The various possibilities are shown in figure 3–4. Panels A through C in figure 3–4 illustrate the three outcomes possible in the example. Panel D in figure 3–4 suggests a fourth possibility, that the second-best function $\overline{B}(A)$ may have several local optima. This suggests that if marginal adjustments are made to $A$, it might be possible to be trapped in a suboptimal position $\overline{N}_2$, rather than the global second-best $\overline{N}_1$.

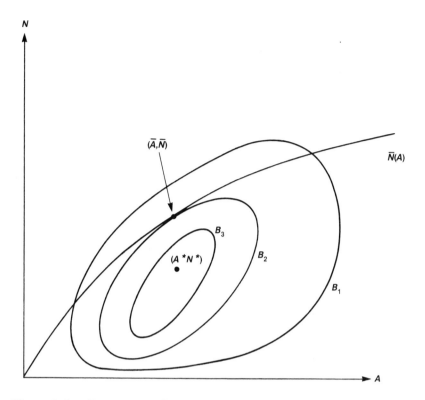

**Figure 3-3.** Comparison of attendance and area under two management regimes

To conclude this section, let us go back to the assumption that everyone in the society is the same, with identical demand and congestion cost functions. Under this assumption, the manner in which visitation is restricted does not affect the expected benefits from using the site or the optimum size of the facility. As noted above, however, the allocation mechanism does affect the distribution of benefits.

When tastes are heterogeneous, the manner in which access is restricted does matter. The most efficient policy is to charge a fee to achieve the optimum attendance, although finding the appropriate fee when tastes are heterogeneous is difficult (Freeman and Haveman, 1977).

Suppose that no fee is charged but access is restricted by refusing admittance when the target level has been reached. Regardless of how the users are chosen, there is no way of ensuring that the ones with the highest willingness to pay are served (Mumy and Hanke, 1975). In fact, Porter (1977) has argued that under some circumstances

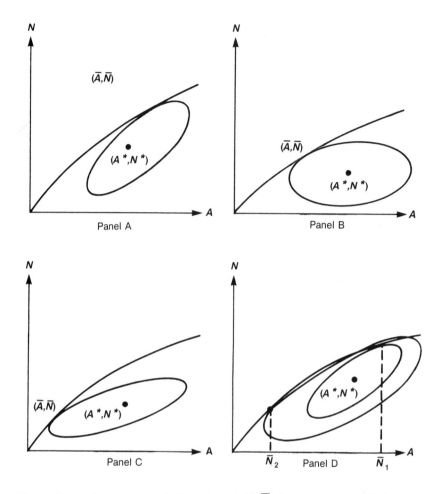

**Figure 3–4.** Comparison of $(A^*, N^*)$ and $(A, \overline{N})$ for various possible benefit functions

it may be less efficient to restrict access by this random process than it is to allow access and incur congestion.

The distribution of benefits is also affected by the allocation mechanism (as well as the quantity allocated), and the most desirable alternative for a given individual depends a great deal on his tastes and circumstances in life. For example, low-income individuals would be more likely to prefer nonmonetary allocation schemes, whereas the rich would prefer rationing based on fees.[5] Because of the differ-

[5]Baden and Stankey (1975) observe that low-income working people might prefer prices to some forms of nonprice allocation, especially those based on first come-first served queueing, because such individuals are most likely to be forced by their work schedules to engage in recreation during the periods of peak use. Students and professional people might have more freedom in this respect.

ence in preferences, Nichols, Smolensky, and Tideman (1971) have suggested that publicly provided goods be made available by way of various allocation schemes. In that way the user could select the facility that most closely accommodates his preferences, abilities, and resources. Anderson and Bonsor (1974) argue that a profit-maximizing monopolist would behave similarly, owning multiple sites and varying the fees charged at the sites to take into account the variation in the tastes and ability to pay to avoid congestion among the public.

## Is the Long-run Cost of Recreation Experiences Increasing?

Earlier sections of this chapter have made clear that the cost of experiences at a site is an increasing function of the number of participants in the short run. That is, if site inputs are held constant, the marginal social cost of a recreation experience is increasing. The marginal cost is found by differentiating $T(N,A)$ in equation (3–12)

$$T_N = NC_N + C(N,A) \tag{3–29}$$

which is an increasing function as long as the congestion cost $C(N,A)$ is increasing in $N$. The average cost function decreases for small values of $N$ as the fixed cost is spread over more and more units, but it begins to increase eventually.

In the long run, the cost is more complicated, but it is not in general true that the cost of recreation experiences is increasing. Complications in measuring the long-run cost of recreation arise in part from the fact, already discussed, that the operator does not necessarily control the output or participation rate. Nonetheless, let us assume initially that the participation rate can be controlled and construct a cost function in the customary manner (for example, Intriligator, 1971, chapter 8). The cost function is constructed in two steps. First, site input(s) $A$ must be selected to minimize the cost of achieving a given level of output $N_0$. This minimum cost for each target output then defines the cost function. To find the cost function for the recreation site, first minimize equation (3–11)

$$T(N,A) = NC(N,A) + K_0(A)$$
$$\text{such that } N \geq N_0 \tag{3–30}$$

The locus of minima $T^*(N_0)$ defines the cost function. Forming the Lagrangian of equation (3–30), we have the following first-order conditions:

$$NC_N + C(N,A) - \lambda = 0 \qquad (3\text{–}31a)$$
$$NC_A + K'(A) = 0 \qquad (3\text{–}31b)$$
$$N = N_0 \qquad (3\text{–}31c)$$

As $N_0$ varies, the functions $\lambda^*(N_0)$ and $A^*(N_0)$ defined by these first-order conditions give the marginal cost and the site inputs required as a function of $N_0$. Differentiating the first-order conditions, we have

$$A^{*\prime}(N_0) = -\frac{C_A + NC_{AN}}{NC_{AA} + K''(A)} \qquad (3\text{–}32)$$

$$\lambda^{*\prime}(N_0) = 2C_N + NC_{NN} - \frac{(NC_{NA} + C_A)^2}{NC_{AA} + K''(A)} \qquad (3\text{–}33)$$

To see how marginal cost varies with $N_0$, we first define two conditions that together give rise to a constant marginal cost function.

(a) The cost of site inputs is constant, or $K_0(A) = wA$
(b) Congestion cost is a linear function of the ratio of $N$ to $A$ (if $A$ is land, then $N/A$ is site density), or $C(N,A) = \theta N/A$

When these functions are substituted into equations (3–32) and (3–33), it turns out that, for all $N_0$,

$$\lambda^{*\prime}(N_0) = 0$$
$$A^{*\prime}(N_0) = (\tfrac{\theta}{w})^{1/2}$$

It is possible to find other combinations of conditions on the components of $T(N,A)$—entirely fortuitous or contrived—that give rise to constant marginal costs. For the most part, however, if either of these conditions does not hold, then marginal cost will either increase or decrease. For some types of recreation, it seems clear that the long-run cost of inputs increases mainly because of a limit on the amount of suitable land. As an extreme example, consider such national parks as Yosemite or the Grand Canyon. Their uniqueness means that it is impossible to add land, and even if other site inputs are added, the fact that one input is fixed means that marginal costs must increase in the long run. Less extreme, but still likely to have increasing long-run costs, is land for skiing and seashore activities. These require suitably located mountains or beachfront land, both of which are finite in supply. Perhaps recreation in general has increasing costs in the long run, especially near urban centers where the long-run opportunity cost of land is rising, as a result of the increase in urban population.

As will be discussed in a subsequent chapter, little is known about congestion cost functions. It may well be that the cost of congestion is indeed accurately represented as a linear function of density in many situations. However, in one important case, scale economies are associated with congestible facilities: if both inputs and outputs are increased by the same factor, congestion costs decrease. This happens when congestion leads to queues and will be discussed in some detail in chapter 5. The point is that we must know both the cost of inputs and the congestion cost to determine the long-run supply curve for recreation experiences at a site.

*four*

# THE AGGREGATE
# RECREATION SUPPLY
# FOR A REGION

In chapter 3, a recreation social cost function for a single recreation area was defined, neglecting travel cost. In this section, we consider the recreation opportunities available within a region, which requires travel cost to be introduced. We first assume that the site facilities are fixed and that access is not limited, and thus that attendance is excessive relative to the optimum, being rationed only by congestion and travel cost.

Both congestion and travel cost figure prominently in the economics of recreation. However, attention to these phenomena has been focused on their implications for benefit estimation. Travel cost, of course, represents a way of estimating willingness to pay for recreation experience and is thus useful in constructing a demand curve (Clawson, 1959). Similarly, interest in congestion has been focused on the effect of site congestion on willingness to pay.

The allocative importance of travel cost and congestion in recreation has received less attention. McConnell and Sutinen (1984) consider congestion as an allocation mechanism in a manner similar in many respects to the analysis presented here, but their focus is on the signals provided to the private recreation supplier, rather than on the construction of indicators of scarcity. McConnell and Sutinen also consider explicitly the implications of enhancing site quality, although they do not consider spatial considerations, as is done here.

## The Single Site Case

Let us begin by assuming that everyone in the society lives in a city located at $x$ and that there is one recreation site located at $y$. Let $t_{xy}$

denote the travel cost from $x$ to $y$, and suppose $R(p)$ is one household's demand curve for recreation experiences, as defined in the preceding section. Thus, in the absence of congestion, the equilibrium frequency of use by the household is given by $R^* = R(p)$, where $p = t_{xy}$. This is the basis of the Clawson-Knetsch travel cost method for estimating demand (Clawson, 1959; Clawson and Knetsch, 1966).

To introduce congestion, let us assume the same upward-sloping congestion cost function $C(N)$ defined in the preceding section (suppressing the argument $A$ for the time being). If $N$ participants are found at the site, the price of an experience to someone located at $x$ is

$$p = t_{xy} + C(N) \qquad\qquad (4\text{--}1)$$

There are two ways of looking at this expression. First, if all participants at the site come from $x$, then equation (4–1) can be viewed as an aggregate supply curve of experiences available to someone at $x$. For each price $p$ there is associated an $N_1$, the number of participants from $x$ who can enjoy the site at the same time. However, this aggregate supply curve is difficult to generalize to a more general spatial pattern of recreation sites and residential location. Also, it requires the assumption that all households have the same congestion cost function. (This is an assumption we make frequently, but only to simplify the exposition.) Alternatively, we can treat $N$ as exogenous and concentrate on the household, in which event equation (4–1) expresses a household supply curve that is horizontal and parameterized by $N$. This approach was used by Deyak and Smith (1978) in an econometric model of recreation demand and supply.

The household's supply and demand functions, together with the attendance at the site, determine the household's equilibrium frequency of use of the site. It is important to note that this equilibrium is partial. The total attendance $N$ at the site is taken to be exogenous: given $N$, the household faces a price $p$, whereupon the demand for recreation by the household is $R(p)$. Figure 4–1 shows how congestion affects both the price and quantity of experiences at the household level. The right-hand quadrant of figure 4–1 relates site attendance and price faced by the household, via equation (4–1). If, for example, site attendance increases from $N_0$ to $N_1$, the effective household price increases from $p_0$ to $p_1$, which reduces the frequency of use from $F_0$ to $F_1$, as shown in the left-hand quadrant of figure 4–1. As noted, the congestion cost function $C(N)$ is specific to the household or individual and is dependent on individual tastes, so that the supply curve for two households at the same location will in general be distinct.

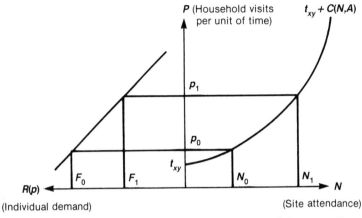

**Figure 4–1.** Effect of congestion on household demand for recreation services

Suppose the city contains $P$ households and let $R_i(p)$ and $C_i(N)$ ($i = 1, 2, \ldots, P$) denote, respectively, the $i$th household's demand curve when the price is $p$ and the congestion cost as a function of the attendance at the site. A general equilibrium solution of recreation participation is then given by the solution to the following $P + 1$ equations in the unknowns $p_1, \ldots, p_p, N$.

$$\sum_{i=1}^{P} R_i(p_i) = N \tag{4–2a}$$

$$p_i = t_{xy} + C_i(N), \ i = 1, \ldots, P \tag{4–2b}$$

Equation (4–2a) gives the equilibrium expected attendance at the recreation site, and equation (4–2b) gives the effective prices faced by each household.

If the demand and congestion cost functions are assumed identical, equation (4–2) can be greatly simplified. In that case, every household faces the same effective price $p$, a market demand curve $D(p) = PR(p)$ can be defined, and the general equilibrium conditions become

$$D(p) = N \tag{4–2a'}$$
$$p = t_{xy} + C(N) \tag{4–2b'}$$

to be solved simultaneously for $p$ and $N$.

The solution—the effective price $p$ and the participation rate $N$—is given by the intersection of the congestion cost curve $(t + C(N))$ and the market demand curve $D(p)$.[1]

[1]Similar reasoning (without introducing congestion costs) is used by Dwyer, Kelly, and Bowes (1977) to show how the travel cost method breaks down for multiple

It should be noted that this outcome is also the solution to a maximization policy. Recall from chapter 3 that the open-access equilibrium maximized benefits to the recreation participants themselves. The participant benefits associated with attendance level $N$ can be written

$$B_p(N) = \int_0^N D^{-1}(q)\,dq - ND^{-1}(N)$$

subject to the constraint $ND^{-1}(N) \geq N(t_{xy} + C(N))$. When $N > 0$, the solution to this problem is given by $D^{-1}(N) = t_{xy} + C(N)$. However, if $D^{-1}(0) < t_{xy}$—that is, the willingness to pay for an uncongested recreation experience is less than the travel cost to the site—then the optimum participation is 0.

Now suppose the population at $x$ can choose between two recreation sites, $A$ and $B$, that provide identical types of experiences, located so that the travel costs are $t_A$ and $t_B$, respectively, with $t_A < t_B$. Suppose the congestion cost functions at each site are $C_A(N_A)$ and $C_B(N_B)$, determined by the area and facilities available at each site. Thus, the effective prices for recreation services facing an individual at $x$ are $p_A = t_A + C_A(N_A)$ and $p_B = t_B + C_B(N_B)$. We assume that an individual, if he participates at all, will participate where the effective price is the lower.

Suppose we distribute individuals one by one over these sites according to this rule. The first person goes to site $A$ because $t_A < t_B$. Subsequent participants continue to go to $A$ until

$$p_A = t_A + C_A(N_A) = t_B \qquad (4\text{--}3)$$

At that point it pays to go to site $B$, and subsequent visits are distributed over sites $A$ and $B$ so that

$$p = t_A + C_A(N_A) = t_B + C_B(N_B) \qquad (4\text{--}4)$$

As in the previous case, these two equations can be solved simultaneously with the demand equation $D(p) = N_A + N_B$ to yield the equilibrium $N$ and $p$, as shown in figure 4-2. Note, again, that the effective prices at the two sites are equal only if both $N_A$ and $N_B$ are nonzero. The attendance at each area, $N_A$ and $N_B$, can be read off the graph. The curve marked $S$ is the horizontal sum of $F_A$ and $F_B$. It looks like and functions as a conventional supply curve.

---

sites. An empirical demonstration of how the effect of increasing inputs is alternated if other sites are available is found in Caulkins, Bishop, and Bouwes (1986).

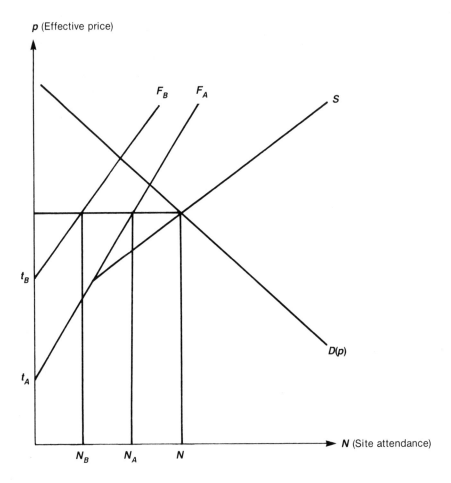

p (Effective price)

$F_B$ $F_A$ $S$

$t_B$

$t_A$

$D(p)$

$N$ (Site attendance)

$N_B$ $N_A$ $N$

**Figure 4–2.** Equilibrium attendance at two recreation sites

A logical problem emerges if we think about how the individual's choice of site is made. Decisions are made not sequentially, as presented above, but more or less simultaneously. In period 1, all participants will go to site $A$ because it is closer. In period 2, the effective price may be lower at $B$ because of the congestion at $A$, in which case all participants go to $B$. The problem, as McConnell and Sutinen (1984) observe, is that congestion costs cannot be posted in advance, but are only known after the event. Thus, it must be assumed that individuals anticipate what other participants will do, and make decisions accordingly. Thus, to determine how and whether an equilibrium is reached, it is important to know how individual expectations are formed. It is possible to model expectation formation and subsequent behavior so that participation and prices oscillate indefinitely, without ever reaching equilibrium. We assume, nonetheless, that this recreation "market" clears, and a stable equilibrium is reached.

Now we turn to the case of two communities, with populations $U_1$ and $U_2$ and travel costs of $t_1$ and $t_2$ to a single recreation site. Again it is assumed that individuals have identical tastes for recreation and congestion. This time, however, the conventional looking demand and supply equilibrium of the previous case is not directly applicable, because the effective prices faced in the two communities are different.

At the equilibrium, prices $p_1$ and $p_2$ faced by residents of the two communities and the total $N$ of expected recreation experiences are given by the solution to the following three equations

$$U_1 D(p_1) + U_2 D(p_2) = N \tag{4-5a}$$
$$p_1 = t_1 + C(N) \tag{4-5b}$$
$$p_2 = t_2 + C(N) \tag{4-5c}$$

The first of these three equations is the demand equation. The next two are the supply equations for individuals in communities one and two, respectively.

The solution of this model is illustrated graphically in figure 4–3. In the northeast quadrant are the "supply curves" $S_1(p)$ and $S_2(p)$, defined as above, for the two communities. The expected demand curves for individuals in each community are plotted in the northwest quadrant. The southwest and southeast quadrants sum the demands and transfer them to the supply axis. Thus, the quantity $N$ of experiences enjoyed translates into prices $p_1$ and $p_2$ paid, respectively, by residents of the two communities. At these prices the quantities demanded are $N_1$ and $N_2$, whose sum is $N$.

This framework can easily be extended to an arbitrary number of residential locations. With $N$ locations the system to be solved contains $n + 1$ equations in the $n$ unknown prices and the total number of experiences $N$.

The responses of the equilibrium solution to shifts in a given parameter value are consistent with intuition and can be determined by differentiating the equations with respect to that parameter and solving the resulting system of linear equations. Thus, to determine the effect of a marginal change in $U_1$, the population in the first community, we solve the following system of equations

$$R(p_1) + U_1 R'(p_1) \frac{\partial p_1}{\partial U_1} + U_2 R'(p_2) \frac{\partial p_2}{\partial U_1} = \frac{\partial N}{\partial U_1}$$

$$\frac{\partial p_1}{\partial U_1} = C'(N) \frac{\partial N}{\partial U_1} = \frac{\partial p_2}{\partial U_1}$$

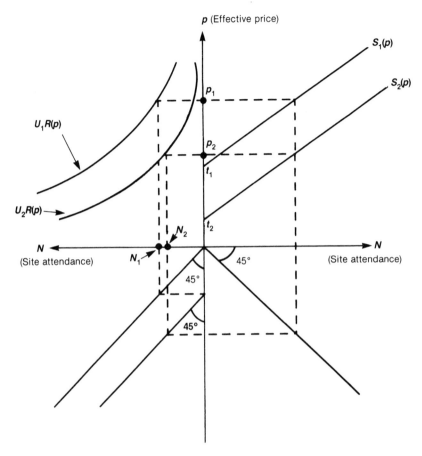

**Figure 4–3.** Equilibrium recreation attendance and effective prices at a recreation site serving two communities

The solution is

$$\frac{\partial p_1}{\partial U_1} = \frac{\partial p_2}{\partial U_1} = \frac{R(p_1)}{\dfrac{1}{C'(N)} - [U_1 R'(p_1) + U_2 R'(p_2)]} > 0$$

$$\frac{\partial N}{\partial U_1} = \frac{1}{C'(N)} \frac{\partial p_1}{\partial U_1} > 0$$

Thus, the effect of a population increase in one of the communities is to increase effective prices in both communities and also increase total visitation $N$. If $N_i = U_iR(p_i)$, it is evident that $\dfrac{\partial N_1}{\partial U_1} > 0$ but $\dfrac{\partial N_2}{\partial U_1} < 0$

Similarly, if the travel cost to one community is changed, comparative statics results can be shown to be the following

$$\frac{\partial p_1}{\partial t_1} = \frac{C'(N)U_2R'(p_2) - 1}{C'(N)[U_1R'(p_1) + U_2R'(p_2)] - 1}$$

so that $0 < \dfrac{\partial p_1}{\partial t_1} < 1$

$$\frac{\partial p_2}{\partial t_1} = \frac{\partial p_1}{\partial t_1} - 1 < 0$$

$$\frac{\partial N}{\partial t_1} = \frac{1}{C'(N)}\left[\frac{\partial p_1}{\partial t_1} - 1\right] < 0$$

Therefore, an increase in travel cost for one community increases the effective price in that community but reduces it in the other community. Total participation declines, although participation in the second community increases.

An increase in site inputs also has a predictable effect on effective prices and number of participants. If we put the site input parameter $A$ back in the congestion cost function and solve the equation system (4–5), we can derive $N$, $p_1$, and $p_2$ as functions of $A$. We find, for example, that

$$\frac{\partial N}{\partial A} = \frac{[U_1R'(p_1) + U_2R'(p_2)]C_A}{1 - [U_1R'(p_1) + U_2R'(p_2)]C_N} > 0$$

Similarly, $\partial p_1/\partial A$ and $\partial p_2/\partial A$ are both negative.


## The Many-Many Case


When there is more than one recreation site and more than one residential location, the problem is more complex and does not admit to a single diagrammatic solution. Suppose that we have $m$ communities or residential zones and $n$ recreation sites, and let

$$N_{ij} = \text{participation at the } j\text{th site by residents of the } i\text{th community}$$

$$N_{i.} = \Sigma_j N_{ij}; \; N_{.j} = \Sigma_i N_{ij}$$

$$t_{ij} = \text{travel cost from the } i\text{th community to the } j\text{th site}$$

$$C^j(N_{.j}, A_i) = \text{congestion cost function at the } j\text{th recreation area}$$

$$D_i(p) = \text{(uncongested) demand curve for recreation experiences in the } i\text{th community}$$

As noted above, the equilibrium solution $\{N_{ij}\}$ maximizes benefits to the participants. The participant benefit function is

$$B_p(N_{11}, \; ..., \; N_{mn}) = \sum_{i=1}^{M} \int_0^{N_{i.}} D_i^{-1}(q_i)\,dq_i - N_{i.}\, D_i^{-1}(N_i) \qquad (4\text{--}6)$$

subject to constraints

$$N_{ij}D_i^{-1}(N_{i.}) \geq N_{ij}\,(t_{ij} + C^j(N_{.j}, A_j)) \qquad (4\text{--}7)$$

For a given community $i$,

$$D_i^{-1}(N_{i.}) = t_{ij} + C^j(N_{.j}, A_j) \qquad (4\text{--}7a)$$

for all $j$ such that $N_{ij} \neq 0$. $p_i = D_i^{-1}(N_{i.})$ is the effective price of a recreation experience for the inhabitants of the community. At equilibrium, the price in community $i$ of using any particular recreation site is the same, or else that site is not visited by residents of that community. If site $j$ is not visited (that is, $N_{ij} = 0$), then $p_i < t_{ij} + C^j(N_{.j}, A_j)$; otherwise net benefits could be increased by increasing $N_{ij}$. If $N_{ij} = 0$ for all $j$, then we must have $D_i^{-1}(0) < t_{ij} + C_j(N_{.j}, A_j)$, thus pricing community $i$ out of the recreation market.

## Quality

As noted, the results of the preceding section assume that the quality of the recreation areas is constant. But inasmuch as recreation areas differ greatly in their quality, that assumption severely limits the applicability of the results. Fortunately, the restriction of constant quality can easily be removed. Suppose the quality index for site $j$ is $Q_j$, and recall from chapter 3 how quality affects the effective price of a recreation experience. If $p(Q_j)$ is the quality adjustment to effective price, as in equation (3–7), the effective price corresponding to equation (4–7a) is

$$p_i = D_i^{-1}(N_{i\cdot}) = t_{ij} + C^j(N_{\cdot j}, A_j) + p(Q_j) \qquad (4\text{-}8)$$

for all $j$ with $N_{ij} \neq 0$.

## Empirical Determination of Effective Price

As it happens, most of the $N_{ij}$ in the solution to equation (4–6) and equation (4–8) must equal zero. Suppose all $N_{ij}$ are nonzero. Then equation (4–8) becomes a system of $mn$ equations in $mn$ unknowns. But these equations are such that

$$C_1(N_{\cdot 1}) - C_2(N_{\cdot 2}) = t_{12} - t_{11} = t_{22} - t_{21} = \cdots = t_{m2} - t_{m1}$$
$$\vdots$$
$$C_1(N_{\cdot 1}) - C_n(N_{\cdot n}) = t_{1n} - t_{11} = t_{2n} - t_{21} = \cdots = t_{mn} - t_{m1}$$

But since $t_{12} - t_{11} = t_{22} - t_{21}$, and so forth, will only be true for each of the $(n - 1)$ equations in equation (4–7), we lose $(m - 1)$ degrees of freedom, so that the rank of this set of equations is at most

$$mn - (m - 1)(n - 1) = m + n - 1$$

For example, if there are thirty residential locations and three recreation sites in the region, then the maximum number of nonzero $N_{ij}$ is thirty-two. Assuming that participants at each of the thirty locations will participate somewhere, that leaves at most two residential locations that visit more than one site. Therefore, if the number of recreation areas is small relative to the number of communities, we shall find a high degree of market segmentation, with participants in most communities visiting only one recreation area. This conclusion also holds for heterogeneous sites after adjustment for quality differences, although clearly it would no longer be valid for sites differing in the type of recreation offered.

The fact that the theory places so many additional constraints on the equilibrium pattern of attendance suggests a heuristic method for empirically determining the effective prices in a region. Suppose we have the attendance $N_{ij}$ and estimates of travel cost $t_{ij}$ from the $m$ residential origins to the $n$ recreation sites, as well as the population $U_i$ of the $i$th residential zone. Now let us assume that demand is linear, that is, $D_i(p) = U_i R(p) = U_i(a - b_p)$ and that congestion is linearly related to density, that is,

$$C(N_{\cdot j}, A_j) = \theta N_{\cdot j}/A_j$$

For given values of the parameters $a, b, \theta$, the model given by equations (4–6) and (4–7) can be solved to yield theoretical estimates of attendance $N_{ij} (a, b, \theta)$. Now we let

$$S(a, b, \theta) = \sum_{i,j} (N_{ij}(a, b, \theta) - n_{ij})^2 \qquad (4-9)$$

and seek the parameters $(a, b, \theta)$ that minimize $S$. That is, we minimize the sum of squared differences between the theoretically implied attendance and the actual attendance levels. The associated effective prices in each community are then given by equation (4–7a).

This approach can be extended to handle differences in site quality. For example, suppose site quality is thought to be determined entirely by the water quality parameter $W$, say $Q = \eta W$, and suppose equation (4–8) holds. Now for $N_{ij} > 0$, the constraints (4–6a) are

$$D_i^{-1}(N_i.) = t_{ij} + \theta N_{ij}/A_{zi} + \eta W_j$$

If origin-destination pairs $N_{ij}$ are observed, we can solve equation (4–9) for the four parameters $(a, b, \theta, \eta)$. It is understood that the quality variable is expressed relative to a reference quality level assigned the value zero. The estimated demand curve is then the zero-quality demand, just as it is the zero congestion demand.

## "Price" Versus "Cost"

As defined in this chapter, the effective price of a recreation experience indicates what the individual must sacrifice in order to enjoy a recreation experience. For persons at different locations, this price will, in general, be different, because the travel costs to the recreation areas will differ. Prices will also vary with with the type of experience and its quality, but these complications do not affect the basic argument.

Let us close this discussion by considering the relationship between this price and the social cost of producing a recreation experience, defined in chapter 3.

In a perfectly competitive situation the price of a good equals its marginal cost. Thus, if recreation were a good traded under perfect competition, the price to the user would equal the marginal social cost of the experience at the site, plus the travel cost. But, prices diverge from marginal costs for various reasons. If the seller has market power, he can maintain a selling price above marginal cost. If external costs are inflicted on third parties during either production or consumption, then price can be less than marginal cost. Prices can also be exceeded by costs if consumption or production of the good is subsidized. Each of these elements—market power, externalities, and subsidies—can affect the provision and pricing of outdoor recreation. As we shall see, subsidization is the most important of these factors.

As noted earlier, the entry fee to most publicly owned recreation facilities in the country is either zero or nominal. Generally, these fees fail to cover even operating costs. According to Gibbs and Reeds (1982), for example, user fees cover only 4 to 8 percent of the annual cost of operating the campgrounds of the Forest Service. Therefore, the effective price of an outdoor recreation experience, as defined in this section, is likely to differ from the social cost of providing the experience. In this respect, recreation is similar to other natural resource products, among which market failure and government intervention are commonplace (examples include water from federally built reservoirs and timber from federal lands).

Whether the subsidization of outdoor recreation in this manner is a good idea is, of course, an important policy question, but one that is beyond the scope of this study. The point of this section is merely to point out that the effective price of a recreation experience is not necessarily a good indicator of its cost. It is, however, a good indicator of the availability of recreation opportunities, subsidized or not.

*five*

# CONGESTION AS A COST

The central argument of this study has been that congestion and travel cost are central elements of the cost of a recreation experience. Treatment of travel cost in this manner is straightforward and conventional. There is less precedent, however, for treating congestion as a cost.

This chapter will examine the nature of recreation congestion and how such congestion imposes costs on the user. By definition, congestion is the loss of satisfaction experienced by the user of a facility as a result of the presence of other users. As users are added at a facility at which site resources are fixed, the satisfaction experienced by each user declines. Congestion is, first and foremost, a peak load problem. Recreation sites typically enjoy peaks during a particular season—usually the summer—and weekly peaks on weekends and holidays. Peaks generally occur for two reasons. First, leisure time is more costly at some times than others, because households are bound by work or school schedules. Thus, weekend recreation is "cheaper" than midweek recreation for those working a standard work week, and summer recreation is cheaper than recreation in other seasons for households with children in school. Second, the quality—and sometimes the feasibility—of most recreation experiences is seasonally determined. Thus, skiing households must take vacations during the winter or early spring, even though the summer might be more convenient.

The timing and magnitude of peak periods are highly site-specific. Two new urban parks in the national park system provide an arresting example. At both the Gateway National Recreation Area in New York City and the Golden Gate National Recreation Area in San Francisco, monthly peaks occur in the summer. At Gateway, however, 68 percent of total annual visits occur in the summer, while at Golden Gate the comparable figure is only 35 percent. As a result of the more evenly distributed pattern of use at Golden Gate, comparable facilities there could provide twice the number of experiences as at Gateway.

Congestion plays an important part in the effective price of a recreation experience; in order to construct a cost function at a specific site, considerable detail is needed on the manner in which congestion costs arise and how such costs are affected by the land area and by facilities at the site. To that end we distinguish between two types of congestion: crowding and queueing. Crowding occurs when no restriction is placed on the number of users at a site. When the number of users is limited, queueing may occur.

One reason for distinguishing between crowding and queueing is that, for the former, no theory exists that is directly useful in determining how recreation users are quantitatively affected by crowded conditions. Empirical research by sociologists and psychologists has uncovered numerous variables other than space that influence feelings of crowdedness, but no quantitative understanding has been achieved. Quantitative results have been produced in some economic studies, in which willingness to pay to avoid crowds has been estimated empirically. However, the economic research has concentrated on physical density, and generally has not considered variables other than space that may affect perceptions of crowding. In contrast, an extensive theory of queueing has been developed in operations research (see, for example, Saaty, 1961). This theory has seldom been applied in economics in general; it has never been used in recreation economics. Queueing is also distinguished from crowding for empirical reasons. Ordinarily, queues can be observed. Crowdedness, on the other hand, is less easily observable, since perceptions of crowdedness are subjective.

# Crowding

Originally, crowding at recreation sites was viewed primarily as a relationship between user density and user satisfaction. This relationship is called the "social carrying capacity" model. This remains the model of congestion costs usually employed by economists in empirical work. In the sociological literature, the social carrying capacity is defined to be the level of use supportable at a given site without causing "excessive" damage to either the physical environment or the users' experience (Lime, 1979). Beginning with Stankey (1973), numerous studies have surveyed recreation users (mainly in wilderness areas and on rivers) and found a relationship between hypothetical density levels and perceived satisfaction. That is, as the number of hypothetical encounters with other parties increases, more and more users reported that they would not enjoy the experience. More recently, Manning and Ciali (1980) found a similar relationship between hypothetical encounters and user satisfaction on rivers in Vermont, but they found no relationship between *actual* encounters and user satisfaction. Similar results have been reported by Bultena and coauthors (1979) and others.

The discrepancy between hypothetical and actual results has called into question the meaning and measurement of recreation satisfaction (Dorfman, 1979). Another difficulty with this research is that these studies necessarily rely on self-reporting of contacts by users, and it has been shown that users are not accurate reporters of contacts (Shelby and Colvin, 1982). The discrepancy has also raised doubts about the social carrying capacity model itself, because that model assumes a direct effect of density on satisfaction (Gramann, 1982). Apparently, user satisfaction is affected not by crowding directly but by *perceptions* of crowding, and user density is only one of several variables affecting perceptions of crowding. Recent crowding research has distinguished between "expectational" and "situational" variables, and has generally found that the discrepancy between expected and actual conditions explains more of the variation in user satisfaction than the actual conditions themselves (Heberlein and coauthors, 1979). Thus, recreation users at remote sites can experience congestion-related losses even though the site would not, by any conventional definition, be considered crowded. In contrast, other recreation participants sometimes seek crowds. College students converge on Fort Lauderdale in the spring seeking other students. And rangers of the National Park Service have often observed what appears to be crowd-seeking behavior among campers at developed areas.[1] To some extent, this may be true of many forms of recreation, which are, after all, social activities.

Thus, the discrepancy between expectations and actual experience is an important determinant of satisfaction. This suggests the potential benefits of aggressively disseminating information about congestion at recreation sites, not only to bring expectations more in line with actual conditions, but also to bring underutilized sites to the attention of participants.

Site conditions are also important, and at least one recent study has found, in contrast to the findings of Heberlein and coauthors, (1979), that actual site conditions may often be more important than expectations (Hammitt, McDonald, and Noe, 1984).

---

[1]Interview by author with Bart Truesdell, Chief Ranger, George Washington Parkway, McLean, Virginia, April 22, 1985.

Even at the site, however, factors other than simple density are important. For one thing, a user is as likely to be affected by the behavior and activities of others as by numbers alone. A user subject to boisterous or antisocial behavior by others is more likely to notice and feel adversely affected by congestion (Gramann and Burdge, 1981). Signs of such behavior might include litter, theft, excessive pollution, or damaged facilities. Even when behavior is not a problem, activities that are mutually incompatible can engender perceptions of crowding among some users. Thus, sailboat users or canoeists may feel more crowding from powerboats than from other sailboats, and tent campers may feel more adversely affected by recreational vehicles than by other tent campers (Lucas, 1967; Driver and Bassett, 1975). The impact of any such incompatibility is likely to be asymmetric: canoeists are far more likely to feel adversely affected by powerboats than the other way around.

These are commonsense notions that suggest that recreation congestion might be relieved by means other than adding land to reduce density. For example, better policing of users or segregation of some users may reduce congestion without affecting user density. At least for some sites, management of a recreation site as a multiple use facility is questionable.

Crowding can also be experienced by recreationists in competition for limited facilities. In one of the few studies of this issue, one fourth of campers at a backcountry campground complained of crowding due to insufficient picnic tables, latrines, and fire pits (Womble and Studebaker, 1981). It is perhaps obvious that equipment use conflicts engender feelings of crowdedness. What is less obvious is the extent of the effect in quantitative terms. Still less clear is how users might value less crowded conditions. The lack of quantification and valuation afflicts all the sociological and psychological research cited above. Although this research has contributed to understanding the "anatomy" of crowding, it is of little direct use in building congestion cost functions.

In contrast, the work of economists on congestion has been explicitly directed at measuring these effects. With one important exception, however, a "black box" approach has been taken, in which congestion effects are estimated for a site, but not tied to any particular site characteristics (except land area).

The exception is found in the body of work on wilderness recreation conducted by John Krutilla and his colleagues over a decade ago. In a study of the Spanish Peaks Wilderness Area in Montana, Cicchetti and Smith (1973) established that a wilderness user's willingness to pay was closely tied to the number of encounters with other groups during his stay in the wilderness. Furthermore, the circumstances of the encounter were also likely to be important (that is, whether it occurred on the trail or in camp, or at the head of the trail or in the middle of the trip). A wilderness travel simulator based on the Spanish Peaks research was developed (Smith and Krutilla, 1976) and applied to the Desolation Wilderness in California (Schechter and Lucas, 1978) and rafting on the Green River in Colorado (Lime, Anderson, and McCool, 1978). The simulation model allowed consideration of policies other than the designation of additional wilderness designed to reduce the number of encounters (and hence the cost of congestion), such as the construction of new trails, or restrictions on the number of groups. Thus, the simulator allows explicit comparison of restricted access and free access, among others.

Unfortunately, nothing of this sort exists for nonwilderness recreation, which, of course, accounts for the overwhelming bulk of recreation activity. Two excellent empirical studies of crowding at developed recreation sites are noteworthy. McConnell (1977) estimated the willingness to pay to avoid congestion on a beach in Rhode Island, finding that, on average, an extra one hundred persons per acre reduces willingness to pay by 25 percent. Walsh, Miller, and Gilliam (1983) investigated two forms of congestion at three ski areas in Colorado. They estimated that the average skier at the areas studied would be willing to pay 27 to 34 cents for each one-minute reduction in lift line wait, and 9 to 22 cents to reduce crowding on the slopes by one person per acre. In addition, this study developed an empirical relationship between the visitation rate, in skiers per day, and both lift line wait and slope crowdedness.

Neither of these studies looked closely at the on-site opportunities of reducing crowding, and for both crowding was measured by persons per acre, which perforce allows crowding to be alleviated only by adding land inputs. It may be that for the recreation resources they examined—beaches and ski slopes—the only inputs that reduce crowding are more beaches and more slopes.

Both studies elicited willingness to pay by means of direct questioning of participants regarding their valuation of hypothetical conditions. In view of the disparity between actual and hypothetical satisfaction found at wilderness areas, this technique is suspect. Nonetheless, valuation by indirect methods is difficult, because visitors' dissatisfaction with crowded conditions is expressed by their staying away. The problem is exemplified by Yogi Berra's comment, "Nobody goes to that bar anymore; it's too crowded."

Two studies of the effects of congestion using indirect methods deserve mention. The first is Deyak and Smith (1978), who use an aggregate measure of congestion, namely average use per acre at National Forest Wilderness and Primitive Areas, by state. The second is the paper by Brown and Mendelsohn (1984), in which they describe the "hedonic travel cost method," a procedure that allows them to estimate not only the willingness to pay to avoid congestion, but other site attributes as well. In their application of this technique to fishing in Washington State, scenery and fish density were examined. Briefly, the technique consists of two steps: in the first, the price of each characteristic was estimated by regressing the distance traveled by each participant in the sample on the level of the attribute reported at that participant's destination. In the second step, inverse demand equations for each attribute were estimated in a two-stage least squares regression of the attribute price, estimated in step one, on the reported quantities of the attributes at each site, together with the fishermen's personal characteristics, such as fishing experience and income.

The data for this study were obtained from a survey of fishermen in Washington, in which the participants identified both their residential location and fishing sites. To obtain the congestion variable, the authors asked each respondent to make a subjective evaluation—on a scale from 1 (worst) to 10 (best)—of crowdedness at each site. These evaluations were then averaged over each site. Unfortunately, these subjective evaluations were not associated with any objective measure of congestion. This means that the results can be used to determine the effective price to a fisherman at a particular residential location, given existing congestion levels, not to determine the ability of the fishing sites to supply fishing experiences.

The hedonic travel cost method has recently been the subject of a searching critique by Smith and Kaoru (1985), who point out that results are substantially affected by the assumptions and analytical procedures that must be adopted. For example, the estimated prices of site attributes are often negative. Should observations with negative attribute prices be included, as Brown and Mendelsohn (1984) do, or excluded, as Bockstael, Hanemann, and Kling (1985) do in another application of the hedonic travel cost approach? In particular, using data collected at recreation sites near Pittsburgh, Smith and Kaoru show that the estimated attribute prices are often negative. They show that the results vary considerably depending on how such negative prices are handled, and that results are greatly affected by the way quantities of site characteristics are defined. They conclude that these difficulties "cast doubt on the ability to develop reasonably plausible and robust estimates of the constituent elements of the model" (p. 25).

Let us return briefly to the observation that crowding generally affects the quality of an experience, but not its cost. The advantage of defining congestion cost in this way is that it converts a quality difference—something difficult to deal with analytically—into something that can be measured, at least in principle. But projecting quality differences onto the price axis via willingness to pay also has disadvantages. For example, consider an individual's choice between two recreation areas, one nearby but highly congested, and the other with lower congestion levels but farther away. If congestion were nothing more than an element of the price of the experience, then the individual would have an absolute preference for one site over the other. That is, one site—the one with the lower sum of travel and congestion costs—would always be chosen, just as one would always choose to pay the lower of two prices offered for identical goods. But it would not be surprising to find both areas being visited by this individual. In effect, the experiences at the two sites are not regarded as identical goods at different prices, but as different goods and the individual might wish to purchase some of each.

Thus, treating congestion as a cost is a compromise with reality. But it is a compromise that must be made if something useful is to be said about recreation resources in the aggregate. Indeed, the whole point of this exercise is to aggregate dissimilar things in a meaningful way. That does not mean that wilderness areas and developed campgrounds ought to be classified together, the only difference between them being a choice between a remote, uncrowded site and an accessible but crowded site. A middle ground has to be found between that position and one that regards each site as unique.

## Queueing

In contrast to crowding, queueing congestion represents a true cost, because while waiting for service the individual could have been doing something else. By the same token, quality is generally unaffected. A tennis player waiting for an open court may not enjoy the wait, but once on the court, the quality of the experience is probably little different from what it would have been had no one else been around.

This sort of congestion results when the number of users at any one time is limited by custom or by the rules of the game. For example, a playground basketball court is generally used by no more than ten participants at a time. It is likely that queueing develops in such situations because the cost of crowding is so high that it reduces the value of the experience to zero. For example, tennis would be no fun if anyone who showed up could use the court regardless of whether play by others was already in progress.

A park operator may choose to substitute queueing for crowding, especially if the crowding is severe enough to destroy the value of the experience. Indeed, some might even regard queueing as a way of dealing with congestion rather than a form of congestion. That is, if a facility is crowded, the operator may put a limit on attendance, and those waiting to get in may form a queue. Usually such a limit will be set where some crowding is experienced, so that a user may have to endure both queueing and crowding. Thus, at the National Air and Space Museum, summer crowds are sometimes so large that tourists are allowed to enter the (still crowded) building only when someone else leaves.

The properties of queues are the subject of a vast literature in operations research, although these methods have been used only infrequently in economics (see, for example, DeVany, 1976; DeVany and Frey, 1982; Naor, 1969; and Mulligan, 1983). Elementary results from queueing theory can be used to quantify congestion effects. Whether this is a step toward a congestion cost function (that is, whether the quantitative measurements of congestion can be valued) is a question to which we shall return at the end of the chapter.

Let us examine two recreation situations that give rise to very different manifestations of congestion.

## *Example 1: A Tennis Court*

A tennis court "produces" recreation opportunities at the rate of one hour of court time per hour for, say, the twelve hours of daylight during which play is possible. If a pair of tennis players arrived once an hour and played for an hour, the court would produce twelve hours of tennis per day, with no waiting. However, the arrival times are stochastic, and it is likely that a pair of players will arrive while others are still playing and will have to wait. At other times the court may stand idle. If arrivals are stochastic the theoretical maximum capacity of the facility cannot be reached, and can only be approached as the average waiting times of the participants increase. By converting this waiting into a cost, we arrive at a cost function for the tennis court.

Suppose pairs of tennis players arrive at random times at a rate of $\lambda$ pairs per hour, wait in line for their turn on the court, and then play for exactly one hour. By "random" we mean that the probability of an arrival of a pair of players in a given time interval is the same, regardless of the arrival time of the previous pair. This assumption leads to an exponential distribution of interarrival times (that is, time between successive arrivals), so that the density function for the time of next arrival is

$$f(t) = \lambda e^{-\lambda t}$$

The interarrival time distribution is a fundamental part of the queueing problem. Also crucial are the service time distribution and the number of servers. In the current context, a service time is simply the duration of play for a waiting pair; the number of servers is the number of courts. Because we have assumed only one court, we have a single-server queue. As noted, our interest is in the steady state conditions, the expected waiting time for a pair entering in the queue. For a single-server queue with an exponential arrival distribution with parameter $\lambda$ and an arbitrary service time distribution $t$ with mean $1/\mu$, the expected waiting time $EW$ is given by the Pollaczek-Khintchine formula (Saaty, 1961, p. 40)

$$EW = \frac{\rho^2 + \lambda^2 \, \text{Var}(t)}{2\lambda(1 - \rho)} \tag{5-1}$$

where $\rho = \lambda/\mu$ is called the *traffic intensity parameter*. (This formula is defined only for $\rho < 1$. If $\rho \geq 1$, the system never reaches steady state and the expected waiting time does not exist. In practical terms this simply states that the tennis court cannot possibly produce more than one hour's play per hour.)

Because we have assumed a constant service time $\text{Var}(t) = 0$, equation (5-1) reduces to

$$EW = \frac{\lambda}{2\mu(\mu - \lambda)} = \frac{\lambda}{2(1 - \lambda)} \tag{5-2}$$

since $\mu = 1$. Let $V$ denote the (common) value of time for all participants, and let $C(\lambda, k)$ denote the congestion cost function for a participant, where $k$ is the number of tennis courts at the site. From equation (5-2),

$$C(\lambda, 1) = \frac{\lambda}{2(1 - \lambda)} \tag{5-3}$$

Suppose it is desired to reduce the expected waiting time, and therefore the congestion cost. One way of doing this is by increasing $\mu$, by forcing a faster turnover on the court. If $\mu = 2$, for example, play would be limited to half an hour. However, by reducing playing time for each participant the operator reduces satisfaction as well.

The only meaningful way to reduce waiting time is to increase capacity—by building new courts or by enhancing the availability of the existing court (for example, by installing lights). Suppose a new court is to be installed. Should the second court be located adjacent to the first court, or at a new location? A new location might seem preferable because travel cost is likely to be lower (and cannot be higher). The situation is not quite so simple, however, because the location of two courts in one location offers efficiencies that cannot be obtained by locating the courts at distinct locations.

First, suppose the new court is built at the new location. If the combined arrival rate at the two courts is $\lambda$, then the arrival rate is $a\lambda$ at one court and $(1 - a)\lambda$ at the other ($0 \le a \le 1$). The congestion costs at the two courts are, respectively, $C(a\lambda, 1)$ and $C((1 - a)\lambda, 1)$, with the average congestion costs for this system being the weighted average of the costs at each court (the weights, of course, being $a$ and $(1 - a)$).

It is routine, but tedious, to show that the average congestion costs are minimized when $a = \frac{1}{2}$, in which case the average congestion costs are

$$C = C(1/2\lambda, 1) = \frac{(\lambda/2)V}{2(1 - \lambda/2)} = \frac{\lambda V}{2(2 - \lambda)} \tag{5-4}$$

Of course, it may not be easy to locate the second court so that $a = \frac{1}{2}$, especially if the first court is at an optimum location (for a single court). But equation (5-4) represents the minimum congestion costs in this situation.

Now contrast this case to the one in which a court is added next to the existing court. With both courts at the same location, those waiting in line play on either court as it becomes free. This is a multiple-server queue, the solution to which can be quite complicated. However, we can easily find an upper bound to the expected waiting time by finding a single-server queue with a greater expected waiting time and then applying the Pollaczek-Khintchine formula (equation (5-1)).

Let $\overline{W}(\lambda, \mu, k)$ denote the expected waiting time in a queue characterized by exponential interarrival times with arrival rate $\lambda$, a constant service frequency $\mu$, and $k$ servers. If $p_n$ denotes the probability that $n$ users are in the system (both in the queue and in service), then

$$\overline{W}(\lambda, \mu, k) = \sum_{n=0}^{\infty} \overline{W}(\lambda, \mu, k|n)p_n$$

where $\overline{W}(\lambda, \mu, k|n)$ is the expected waiting time of a unit entering the queue given that there are already $n$ users in the system.

If $n < k$, we have $\overline{W}(\lambda, \mu, k|n) = 0$, because at least one server is unoccupied. If $n \ge k$, then $\overline{W}(\lambda, \mu, k|n) = \overline{W}(\lambda, k\mu, 1|n)$. Thus, for example, when both servers are occupied, the waiting line for two servers is the same as for one server working twice as fast. Therefore, $\overline{W}(\lambda, \mu, k|n) \le \overline{W}(\lambda, k\mu, 1|n)$ for all $n$, and hence $\overline{W}(\lambda, \mu, k) \le \overline{W}(\lambda, k\mu, 1)$.

Applying this reasoning to the two tennis courts, we find, using equation (5-2), that

$$\overline{W}(\lambda,1,2) \le \overline{W}(\lambda,2,1) = \frac{\lambda}{2(2)(2-\lambda)} = \frac{\lambda}{4(2-\lambda)} \qquad (5\text{-}5)$$

Comparing equations (5–4) and (5–5) it is evident that the waiting time for two courts located together is less than half the waiting time of two courts located at distinct locations.

This result obtains because two adjacent courts are much more efficient at providing services than two separate courts. The increase in efficiency comes from the fact that, for adjacent courts, whenever there are two or more pairs in the system both courts must be occupied. However, with separate courts it is possible for one court to be empty no matter how many pairs are in the system. There are, thus, scale economies associated with the addition of capacity at an existing site rather than a new site. These economies are related entirely to the stochasticity of arrivals and departures.

### Example 2: A Campground with "No Vacancy"

As a second application of queueing theory to the characterization of congestion cost functions, consider a campground with $N$ campsites. Again we assume exponential arrival times, and this time we have exponentially distributed departures as well. Thus, in the next time interval $\Delta t$ the probability of an arrival is $\lambda(\Delta t)$, and the probability that the occupants of a given campsite will leave is $\mu(\Delta t)$. If $n$ campsites are occupied, therefore, the probability of a departure in the next $\Delta t$ is $n\mu(\Delta t)$. Finally, we assume that no waiting line forms; a party arriving when all campsites are filled is turned away.

These assumptions define a Markov process with states $S_0, \ldots, S_N$, with states defined by the number of campsites occupied. The matrix of transition probabilities is thus given by

|       | $S_0$       | $S_1$              | $S_2$           | $\cdot$ | $\cdot$ | $\cdot$ | $S_N$         |
|-------|-------------|--------------------|-----------------|---------|---------|---------|---------------|
| $S_0$ | $(1-\lambda)$ | $\lambda$        |                 |         |         |         |               |
| $S_1$ | $\mu$       | $1-\lambda-\mu$    | $\lambda$       |         |         |         |               |
| $S_2$ | $0$         | $2\mu$             | $1-\lambda-2\mu$ | $\lambda$ |       |         |               |
| $\cdot$ | $\cdot$   |                    |                 |         |         |         |               |
| $\cdot$ | $\cdot$   |                    |                 |         |         |         |               |
| $\cdot$ | $\cdot$   |                    |                 |         |         |         |               |
| $S_N$ | $0$         |                    |                 |         |         | $N\mu$  | $(1-N\mu)$    |

Let $p_0, p_1, ..., p_N$ denote the stationary probabilities of this Markov process. The $\{p_i\}$ satisfy the following system of equations:

$$p_0 = (1 - \lambda) p_0 + \mu p_1$$
$$p_1 = \lambda p_0 + (1 - \lambda - 2\mu) p_1 + 2\mu p_2$$
.

.

.

$$p_{N-1} = \lambda p_{N-2} + (1 - \lambda - N\mu)p_{N-1} + N\mu p_N$$
$$p_0 + p_1 + \cdots + p_N = 1,$$

the solution to which is

$$p_0 = \left[ \sum_{j=0}^{N} \frac{(\rho)^j}{j!} \right]^{-1}$$

$$p_n = \frac{1}{n!} \rho^n p_0 \ (n = 1, 2, ..., N)$$

where $\rho = \lambda/\mu$

This is an example of queueing with balking. In a balked queue, new arrivals refuse or are not permitted to join the queue when the waiting line reaches a certain length (in this case, zero). Unlike the previous example, here there is no restriction on the arrival rate, for with no waiting line there is obviously no possibility of the waiting line growing without bound. As the traffic intensity increases, most new arrivals will be turned away because all campsites are occupied.

The frequency with which new arrivals encounter a "no vacancy" is the balking frequency $B$ and is equal to $\lambda p_N$, a measure of the congestion in the system.

Let $M(p,N)$ denote the expected number of campsites occupied when there are $N$ campsites and the traffic intensity is $\rho$

$$M(\rho,N) = \sum n p_n$$

$I(\rho,N) = N - M(\rho,N)$ is the average number of unoccupied campsites, a measure of the idle capacity in the system. Generally, decreasing the balking frequency $B$ can only be achieved by increasing the average number of unoccupied sites. The tradeoff between disappointed arrivals ($B$) and idle campsites ($I$) is illustrated in figure 5–1, which plots $B$ versus $I$ for $\lambda = 3$, $\mu = 0.5$, and $N=4,5,6,7,8$. As shown, adding campsites reduces the effects of congestion, but at a steadily decreasing rate.

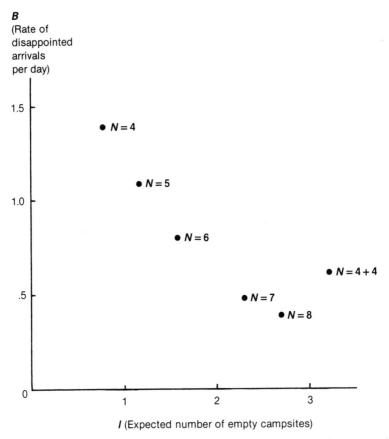

**Figure 5–1.** Tradeoff between idle campsites and disappointed campers for $\lambda = 3$ and $\mu = 0.5$

As in the preceding example, scale economies are associated with locating all the campsites in one spot. Compare, for example, two facilities with four campsites each. Assuming each facility serves half the demand we have the following values for the parameters

$$\lambda = 1.5$$
$$\mu = 0.5$$
$$N = 4$$

which yields

$$B = .309$$
$$I = 4 - 2.38 = 1.62$$

These values for $B$ and $I$ are for each location; to obtain the balking frequency and expected number of idle campsites for the whole system these values must be doubled. They are plotted on figure 5–1 and labeled "$N = 4 + 4$." As shown, with two sites there are, on average, more disappointed campers and more idle campsites than when all campsites are in one location. As before, of course, these economies may be offset by the reduction in participants' average travel cost with two sites.

In these two examples, the effects of congestion are very different, but they make it clear that the effects of congestion can often be quantified. Indeed, methods from queueing theory could also be used to analyze congestion in numerous other recreation contexts, such as roads or trails, boat slips, and ski lifts. Queueing models can be complex and mathematically cumbersome, often giving rise to results that cannot be expressed in closed form. However, virtually any queueing phenomenon can be simulated by Monte Carlo methods as long as the queue discipline and the arrival and service distributions can be identified.

Still, it is not clear that this is a gain. Some might argue that the argument presented above is the equivalent of the farmer who weighs his produce by carefully balancing it against a pile of stones, and then guesses the weight of the stones. Even though nearly everyone would agree that the effects of congestion are undesirable, valuing just how undesirable they are is difficult.

The cost of waiting in line is heavily dependent on the value of time. Following Becker (1965) the standard procedure for valuing time is to use the wage rate, or some multiple of the wage rate (as did Brown and Mendelsohn, 1984). Valuing time at the wage rate implicitly assumes that individuals are able to trade work time for leisure time continuously. But there are indivisibilities and other imperfections in labor markets that make this assumption quite tenuous. Nonetheless, the few studies of the value of waiting time conclude that its value is approximated by the wage rate (Deacon and Sonstelie, 1985), even though the variance is high.

In addition, it should be noted that the cost of waiting time may depend a great deal on particular circumstances. Consider, for example, the wait for a table at a restaurant. Standing in line outside in the cold and sitting comfortably in a cocktail lounge both represent waiting. However, the cost of waiting in these two situations would be greatly different, even though the length of the wait might be the same. Studies of commuters have estimated the value of time spent in a car to be only 20 to 30 percent of the wage, compared to estimates of 70 percent of the wage rate for the value of time waiting for public transit (Bruzelius, 1979).

Even more elusive than the cost of waiting time—and probably even more situation specific—is the cost of encountering a "no vacancy." If there are acceptable alternatives nearby, the cost may be very low. If the alternatives are also filled to capacity, the cost could be high. The potential costliness of being turned away is suggested by the actions people take to avoid it: they make reservations, often with cash deposits, and when that is not possible, they often schedule their travel to increase their chances of being served. For example, an individual might leave work early on Friday, sacrificing wages, to ensure that camping sites will still be available at his or her destination. Indeed, the cost of taking such actions can also be quite properly considered a cost of congestion. Unfortunately, there has been virtually no empirical work on congestion of this sort. Unlike waiting time, there is not even a remote market proxy, such as wages.

It might seem, then, that the contribution of queueing theory to recreation congestion problems is limited, not only by the difficulty of valuation but also by its apparent applicability only to problems of queueing rather than crowding. Appearances to the contrary notwithstanding, methods from queueing theory, including Monte Carlo simulation, can usefully analyze almost any congestion problem in which the effects on individual units in the system can be quantified. Besides waiting time or the probability of being turned away, as in the two examples above, these effects include the number of backcountry encounters. As noted earlier, such encounters have been identified as an undesirable consequence of crowding in wilderness recreation and have been examined via a simulation model.

As a general matter, quantification of an effect is a step forward even when valuation is difficult or impossible. Valuation is usually not necessary to make comparisons across time or space, for example. In fact, measurement without valuation is common in almost every branch of public policy analysis. In environmental policy, for example, air and water quality are measured in terms of air or water contaminants. Those looking for benefits of improving air quality examine the relationship between air quality levels and welfare measures such as morbidity or mortality, the valuation of which is no doubt as difficult as the valuation of congestion effects.

Moreover, valuation of the microeffects of congestion may well be of more use than the current methods of valuing recreation congestion. Recall that such methods attempt to link willingness to pay either to aggregate indicators of congestion, such as site density (for example, McConnell, 1972) or to participants' subjective impressions (for example, Brown and Mendelsohn, 1984). Each of these indicators fails to make the essential link between the number of individuals on the site (or the arrival rate, or rate of use) and the precise manifestations of congestion that reduce willingness to pay. As a result, the estimates of congestion costs are very site specific and therefore impossible to generalize, because one does not know precisely what causes the reduction in satisfaction. By linking arrival rate with waiting time and other consequences that have direct implications on utility, a detailed examination of the effects of congestion using queueing theory may help bridge this gap.

# six

# CONCLUSION

The impetus for this monograph was the concern, expressed in many quarters, over the adequacy of the supply of recreation resources. It has been argued that it is currently impossible to determine whether recreation resource availability is adequate or not, or even whether it is increasing or decreasing, or how different regions compare in the availability of recreation opportunity. Current measures of recreation resource availability, expressed in purely physical terms, are not very informative regarding the ability of those resources to produce valued recreation experiences for participants.

For a marketed commodity, the chief indicator summarizing its scarcity is its price. Price indicators are not available for most outdoor recreation services, of course, because they are publicly provided and are not traded in markets. But we can devise surrogates for prices that indicate what must be given up in order to enjoy a recreation experience.

Under perfect competition, price equals marginal cost. Accordingly, the search for recreation price surrogates began with an examination of the cost of producing a recreation experience at a given site (chapter 3). These costs were shown to be shared between the operator of the site and the recreation participant. For participants, a major element of the cost of a recreation experience is site congestion. Indeed, at many recreation sites, the operator has only fixed costs, and but for congestion, recreation experiences would be available at zero marginal cost. Under those circumstances, recreation would have the nonrivalness property of a public good and questions of resource availability would be much less interesting. Congestion is therefore at the center of the recreation supply issue.

The site operator is usually able to influence the level of congestion costs in the short run by means of the pricing and admission policies at the site. In the long run, congestion costs can be modified by additions of land or capital. Concerns about the adequacy of recreation supply have often been predicated on an assumption that the cost of recreational land will be rising, especially in urban areas, reflecting competition from other uses. But it is shown in chapter 3 that rising land prices do not necessarily imply increasing recreation resource scarcity, because of the substitution possibilities and potential economies of scale. In chapter 5 it is shown that scale economies can arise purely for stochastic reasons.

For individual participants, the costs at the site are only part of the sacrifice required to enjoy a recreation experience; the cost of travel to the site must also be endured. Both congestion cost and travel cost appear prominently in recreation economics, but almost exclusively for recreation benefit estimation. In chapter 4, they are also shown to have an allocative function. In a spatial model of $m$ communities and $n$ recreation areas, recreation participants are shown to distribute themselves so that an equilibrium is achieved in which the attendance from each community to each site is determined by the sum of congestion and travel costs. For any given community, this sum is the same for all sites, and is called the *effective price* because it fulfills the same allocative function as prices in conventional markets. These results are a generalization of the simple intersection of demand and supply to determine prices, for there is no single price but a different price for each location in space. The effective price determination can further be modified to take into account differences in quality among various sites.

Because this effective price is a measure of what the user must give up in order to enjoy the experience, it is a measure of recreation resource availability, much as the price of natural resources can be considered a measure of their scarcity. If these prices could be determined, comparisons of the recreation opportunities over time at the same location could be made. Assuredly, the differences between the price of natural resource commodities, such as minerals or timber, and the price of a recreation experience, as defined here, are obvious. First, because commodities are traded in markets, their prices can be observed from market transactions. In contrast, for most types of recreation, at best only a relatively small part of the cost of the experience is observed directly in a market transaction. The other components of the price must be obtained by other means. A second difference is that commodity prices are the same to all buyers under most circumstances, while the price of a recreation experience is specific to the household or individual, depending on the location or the tastes of the household. A third difference is that while market prices provide both information and incentives to producers, effective prices of recreation provide only information.

Other attributes and limitations of effective prices should be noted. For one thing, insofar as price depends on level of use, prices will be different for peak and off-peak periods. In general, off-peak prices will not include a congestion component. Second, throughout the discussion it has been assumed implicitly that there is an activity called "recreation." But to bring this concept to life it must be acknowledged that many types of recreation exist, and each will have its own effective price. Using prices of individual activities to assess resource availability must be done with care. For example, the state of Florida is notably impoverished when it comes to skiing availability. But it would clearly be unwise to draw welfare implications from this fact, because (a) those who choose to live in Florida are likely to be those who do not value skiing highly, and (b) different forms of outdoor recreation are almost certainly substitutes, at least in part, although very little is known about the marginal rates of substitution.

Another consideration has to do with multiple-day trips. For convenience we have not specified the duration of the recreation "experiences" discussed above. If we take the length of an experience to be a day, then the effective price per day of a two-day trip to a site is less than the effective price of a one-day trip, because only one trip is required. The same consideration applies if several sites are visited on the same trip. This is true here as it is for ordinary commodities: the fact that longer trips are less expensive per unit is simply a volume discount.

It is conceded, therefore, that even under the best of circumstances these prices would be somewhat arbitrary. They will necessarily be based on the estimated cost of congestion at local recreation sites to an average participant, and will not generally reflect the cost of congestion to any particular individual at that location. Nonetheless, even fictitious prices, based on the tastes of the average participant and reported levels of use at recreation sites, would be informative on the availability of recreation resources.

Such prices would have numerous uses, especially in assessing trends in resource availability over time and the distribution of resources in space. In addition, prices of this sort would be very helpful to policy-makers making decisions about the construction and location of new facilities, especially if demand curves were also available. For example, the location of a new facility in a region would have the effect of lowering effective prices throughout the region. If the distribution of price changes could be determined—perhaps by the sort of equilibrium analysis described in chapter 4—then the expected benefits of the new facility could be calculated and compared to its cost. Another use of effective prices would be to determine the value of recreation assets for an expanded set of national accounts. But to calculate effective prices that are credible will clearly require much additional data and analysis.

A vast amount of data on recreation resources already exists. Over the last decade, each state has prepared a Statewide Comprehensive Outdoor Recreation Plan (SCORP) with funds from the now defunct Bureau of Outdoor Recreation in the Department of the Interior. Many of the SCORPs contain detailed recreation inventories at the county level, with data not only on total acreage but on type of acreage as well. The Florida SCORP, for example, contains information on twenty-eight types of facilities, including miles of saltwater beach, hiking and bicycle trails, and number of campsites, picnic tables, tennis courts, golf courses, swimming pools, and playgrounds. These data are disaggregated by multicounty planning region (each region contains about seven counties), and by ownership class (state, federal, county, municipal, or private). Unfortunately, the SCORP does not report on the location or size of individual facilities, making it impossible to determine the average distance that must be traveled by a local resident to get to these sites. Those data must have existed at some point, in order to aggregate up to the regional totals. The problem of data assembly would be formidable, however.

The federal agencies with the most extensive recreation responsibilities (the National Park Service, the Forest Service, the Bureau of Land Management, the U.S. Army Corps of Engineers, and the Bureau of Reclamation) also maintain comprehensive inventories of recreation acreage, with some data bases covering manmade facilities as well. The contents of these data bases have been summarized by Yardas and coauthors (1982).

Participation data are more problematical. Few of these plans report visitation by site, and none matches origins and destinations. For the most part, participation data are obtained by surveying households, and what site-specific information there is is usually an annual average. The Florida SCORP lists average participation rates for each region for twenty-six different activities for both Florida residents and tourists. However, it is impossible to determine where the reported recreation activities take place. Most states do keep visitation records for their state parks, but often only annual figures. Monthly totals, or totals for even shorter intervals may be available, but would in all likelihood have to be extracted from log books or other records kept at the site.

This is true at the federal level as well. The National Park Service, for example, publishes total yearly visits to units in the National Park System, in its annual *National Park Service Statistical Abstract*. Monthly data from each unit are also available. Data for shorter time periods may be obtained only from individual sites. Data on user characteristics, including residential location, are rarely collected.

In short, the data that are collected are not very useful for determining the main components of effective prices, namely congestion costs and travel costs. It is of interest that all of the studies cited in chapter 5 as the most useful calculations of congestion cost (McConnell, 1977; Walsh, Miller, and Gilliam, 1983; Cicchetti and Smith, 1973; and Brown and Mendelsohn, 1984) had to generate their own data rather than rely on existing data bases.

Can anything be done with existing data to compute effective prices? Given enough assumptions, such prices can be computed. One methodology is as follows:

a. Create a county-level data base consisting of total recreation acreage and average annual use for a particular type of recreation, say, nonwilderness camping.

b. Compute the average distance between a resident and a site in the county. Such a calculation would require knowledge of the number of sites, something that could be computed from average site size if not available directly. This step also requires some assumptions about the location of residents and the location of sites. (One approach, for example, would be to locate the county population at the centroid of the county and distribute campsites uniformly.) Make the same calculation for the average distance from residents of each county to nearby counties. (The definition of "nearby" would, of course, have to be determined.)

c. Compute congestion levels for the campsites in each county. This might be done by taking the average annual participation at campsites in the county and converting the annual figure to a peak and off-peak visitation rate.

d. Devise a scheme for converting participation rates into congestion costs. Adoption of use per acre, valued at several alternative values, would be one possible approach.

e. Compute effective prices of recreation for residents of each county, using various values for the per unit cost of travel and the cost of congestion.

This procedure would produce an effective price for various types of recreation experiences for each county in a region. Many questionable assumptions are required, however, and the calculation would reveal nothing about the importance of manmade improvements in reducing congestion costs. If we were to find, however, that the ranking of locations is insensitive to the assumptions regarding travel and congestion cost parameters, we might be able to make qualitative statements about recreation availability. In any event, effective prices would almost certainly be more informative than the rudimentary acres per capita calculations that figure so heavily in macro-level recreation policy analysis.

At this point, then, the most important gap in our knowledge about recreation supply is the cost of congestion. Research on congestion can be carried out in three distinct ways, each of which has its advantages and disadvantages. The first category includes contingent valuation (CV) studies, in which recreation participants are asked directly how much they would be willing to pay for less crowded conditions. Studies of this sort include those by McConnell (1977), Cicchetti and Smith (1973), and Walsh, Miller, and Gilliam (1983), discussed in chapter 5. The advantage of the CV approach is that the phenomenon of interest—congestion in this case—can be isolated and a direct measurement of its value made. Also, a CV study is most likely to provide information on the distribution of willingness to pay to avoid congestion. On the other hand, CV estimates are hypothetical and may therefore be misleading. Also, unless many studies on many different recreation areas are performed, it is difficult to determine the role of specific characteristics in affecting congestion.

The second approach is simulation. As noted, this approach has been applied to wilderness recreation (Smith and Krutilla, 1976) and could profitably be applied to other forms of recreation. For example, a simulation of a developed state park could probe the relationship between the operator's inputs and congestion in the form of queues or crowded facilities, revealing whether a new road, say, or a new camping area would better reduce congestion at given levels of use. The values of such results would be greatly enhanced by better estimates of the value of time.

The third kind of research is the indirect valuation of congestion based on market or behavioral evidence. The great advantage of this "demand revelation" approach is that it depends on what people actually do, rather than what they say they do, and is therefore more credible. On the other hand, people do things for many different reasons, and actions may be misinterpreted. One approach, discussed in chapter 4, is the "hedonic travel cost" method employed by Brown and Mendelsohn (1984). Applying this approach, however, raises several issues that are difficult to resolve on theoretical grounds and that often have a serious effect on the results. Additional research on the hedonic travel cost method may suggest appropriate ways of dealing with these problems.

Research in these three areas will benefit from more fundamental research on the nature of congestion and its effects on satisfaction of the sort now conducted by other social scientists and reported in journals such as *Leisure Sciences* and the *Journal of Leisure Research*. As noted in chapter 5, until now this research has not produced any quantitative results. Perhaps there is an opportunity here for collaborative efforts involving both economists and sociologists or psychologists, with the objective of valuing the consequences of congestion established by the latter.

These theoretical and empirical issues notwithstanding, calculation of effective prices would yield significant insights into the availability of recreation resources. Data from a large number of different sources can be used to produce a rudimentary set of such prices. More defensible estimates will require extensive empirical work, for there is little existing empirical work on which to draw. It is hoped that the concept of effective price will provide a rationale and a useful organizing principle for the completion of such studies.

# REFERENCES

Anderson, F. J., and N. C. Bonsor. 1974. "Allocation, Congestion and the Valuation of Recreational Resources," *Land Economics* vol. 50 (February) pp. 51–57.

Baden, John, and George Stankey. 1975. "Rationing Wilderness Use: Some Administrative and Equity Implications," Intermountain Forest and Range Experiment Station, Missoula, Montana.

Barnett, Harold, and Chandler Morse. 1963. *Scarcity and Growth* (Baltimore, Md., Johns Hopkins University Press for Resources for the Future).

Barney, Gerald O. 1980. *The Global 2000 Report to the President* (Washington, D.C., Council on Environmental Quality).

Baron, Mira, and Mordechai Schechter. 1973. "Simultaneous Determination of Visits to a System of Outdoor Recreation Parks with Capacity Limitations," *Regional and Urban Economics* vol. 3, no. 4, pp. 327–59.

Baumol, William J., and Wallace E. Oates. 1975. *The Theory of Environmental Policy* (Englewood Cliffs, N.J., Prentice Hall, Inc.).

Becker, Gary S. 1965. "A Theory of the Allocation of Time," *Economic Journal* vol. 75 (September ) pp. 493–517.

Bockstael, Nancy E., W. Michael Hanemann, and Catherine Kling. 1985. "Modeling Recreation Demand in a Multiple Site Framework." Paper presented at the AERE Workshop on Recreation Demand Modeling, Boulder, Colorado, May 17–18.

Brown, Gardner, and Robert Mendelsohn. 1984. "The Hedonic Travel Cost Method," *Review of Economics and Statistics* vol. 65, no. 3 (August) pp. 427–33.

Bruzelius, Nils. 1979. *The Value of Travel Time: Theory and Measurement* (London, Croom Helm).

Bultena, G., D. Field, P. Womble, and D. Albrecht. 1979. "Closing the Gates: A Study of Backcounty Use–Limitation at Mount McKinley National

Park." Paper presented at the Second Conference on Scientific Research in the National Parks, San Francisco, November 26–30.

Caulkins, Peter, R. C. Bishop, and N. W. Bouwes. 1986. "Travel Cost for Lake Recreation: A Comparison of Two Methods for Increasing Site Quality and Substitution Effects," *American Journal of Agricultural Economics* vol. 68, no. 2, pp. 291–97.

Cicchetti, Charles J., and V. Kerry Smith. 1973. "Congestion, Quality Deterioration and Optimal Use: Wilderness Recreation in the Spanish Peaks Primitive Area," *Social Science Research* vol. 2 (May) pp. 15–30.

_____, and _____. 1976. *The Costs of Congestion* (Cambridge, Mass., Ballinger Publishing Company).

_____, J. J. Seneca, and P. Davidson. 1969. *The Demand and Supply of Outdoor Recreation* (New Brunswick, N.J., Bureau of Economic Research, Rutgers–The State University).

Clawson, Marion. 1959. "Methods of Measuring Demand for and Value of Outdoor Recreation," Reprint No. 10, Resources for the Future, Washington, D.C., pp. 2–7.

_____. 1984. "Effective Acreage for Outdoor Recreation," *Resources* vol. 78 (Fall).

_____, and J. L. Knetsch. 1966. *Economics of Outdoor Recreation* (Baltimore, Md., Johns Hopkins University Press for Resources for the Future).

_____, and Carlton Van Doren, eds. 1984. *Statistics on Outdoor Recreation* (Washington, D.C., Resources for the Future).

Deacon, Robert T., and Jon Sonstelie. 1985. "Rationing by Waiting and the Value of Time: Results from a Natural Experiment," *Journal of Political Economy* vol. 93, no. 4, pp. 627–47.

De Meza, David, and J. R. Gould. 1985. "Free Access vs. Private Ownership: A Comparison," *Journal of Economic Theory* vol. 36, pp. 387–91.

DeVany, Arthur. 1976. "Uncertainty, Waiting Time and Capacity Utilization: A Stochastic Theory of Product Quality," *Journal of Political Economy* vol. 84, no. 3, pp. 523–41.

_____, and Gail Frey. 1982. "Backlogs and the Value of Excess Capacity in the Steel Industry," *Americal Economic Review* vol. 72, no. 3, pp. 441–51.

Deyak, Timothy A., and V. Kerry Smith. 1978. "Congestion and Participation in Outdoor Recreation: A Household Production Function Approach," *Journal of Environmental Economics and Management* vol. 5, pp. 63–80.

Diamond, Henry L., E. C. Castle, S. Coleman, W. P. Matt, Jr., P. F. Norman, W. K. Reilly, L. S. Rockefeller. 1983. "Outdoor Recreation for America" (Washington, D.C., Resources for the Future).

Dorfman, P. 1979. "Measurement and Meaning of Recreation Satisfaction: A Case Study of Camping," *Environment and Behavior* vol. 8, no. 3, pp. 345–64.

Driver, B., and J. Bassett. 1975. "Defining Conflicts Among River Users: A Case Study of Michigan's Au Sable River," *Naturalist* vol. 26, no. 1, pp. 19–23.

Dwyer, J. F., J. R. Kelly, and M. D. Bowes. 1977. "Improved Procedures for Valuation of the Contribution of Recreation to National Economic Development" (Urbana, Ill., Water Resources Center, University of Illinois).

Fisher, Anthony, and John Krutilla. 1972. "Determination of Optimal Capacity of Resource-Based Recreation Facilities," *National Resources Journal* vol. 12, no. 3 (July) pp. 417–44.

Freeman, A. Myrick III, and Robert H. Haveman. 1977. "Congestion, Quality Deterioration and Heterogeneous Tastes," *Journal of Public Economics* vol. 8, pp. 225–32.

Gibbs, K. C., and F. L. Reeds. 1982. "Estimation and Analysis of Management Opportunities for Developed Recreation in the Northern Region," Final Report submitted to U.S. Forest Service, Oregon State University, Corvallis, Ore.

Gramann, James H. 1982. "Toward a Behavioral Theory of Crowding in Outdoor Recreation: An Evaluation and Synthesis of Research," *Leisure Sciences* vol. 5, no. 2, pp. 109–26.

———, and R. Burdge. 1981. "The Effect of Recreation Goals on Conflict Perception: The Case of Water Skiers and Fishermen," *Journal of Leisure Research* vol. 13, no. 1, pp. 15–27.

Haight, Frank A. 1957. "Queueing with Balking," *Biometrika* vol. 44 (December) pp. 360–69.

Hammitt, William E., Cary D. McDonald, and Frank P. Noe. 1984. "Use Level and Encounters: Important Variables of Perceived Crowding Among Nonspecialized Recreationists," *Journal of Leisure Research* vol. 16, no. 1, pp. 1–8.

Head, John. 1962. "Public Goods and Public Policy," *Public Finance* vol. 17, no. 1, pp. 197–219.

Heberlein, Thomas, G. Alfano, B. Shelby, and J. Vaske. 1979. "Expectations, Preferences, and Feeling Crowded in Recreation Activities." Paper presented to the Annual Meeting of the Rural Sociological Society, Burlington, Vermont, August 23–26.

Intriligator, Michael D. 1971. *Mathematical Optimization and Economic Theory* (Englewood Cliffs, N.J., Prentice Hall, Inc.).

Kelman, Steven. 1981. *What Price Incentives?* (Boston, Masss., Auburn House Publishing Company).

Lancaster, K. J. 1966. "A New Approach to Demand Theory," *Journal of Political Economy* vol. 74, no. 1, pp. 132–57.

Lime, David W. 1979. "Carrying Capacity," *Trends* vol. 16, no. 2, pp. 37–40.

Lime, David W., D. H. Anderson, and S. F. McCool. 1978. "An Application of the Simulator to a River Recreation Setting," in M. Schechter and R. C. Lucas, eds. *Simulation of Recreational Use for Park and Wilderness Management* (Baltimore, Md., Johns Hopkins University Press for Resources for the Future).

Lucas, Robert. 1964. "Wilderness Perception and Use: The Example of the Boundary Waters Canoe Area," *Natural Resources Journal* vol. 3, no. 3, pp. 394–411.

Manning, Robert G., and C. P. Ciali. 1980. "Recreation Density and User Satisfaction: A Further Exploration of the Satisfaction Model," *Journal of Leisure Research* vol. 12, no. 4, pp. 329–45.

McConnell, Kenneth E. 1977. "Congestion and Willingness to Pay: A Study of Beach Use," *Land Economics* vol. 53, no. 2 (May) pp. 185–95.

———, and Jon G. Sutinen. 1984. "An Analysis of Congested Recreation Facilities," *Advances in Applied Micro–Economics* vol. 3, pp. 9–36.

Mills, Edwin S. 1972. *Urban Economics* (Glenview, Ill., Scott, Foresman and Company).

Mulligan, James G. 1983. "The Economies of Massed Reserves," *American Economic Review* vol. 73, pp. 725–34.

Mumy, Gene E., and Steve Hanke. 1975. "Public Investment Criteria for Underpriced Public Projects," *American Economic Review* vol. 65, no. 4, pp. 712–20.

Naor, P. 1969. "The Regulation of Queue Size by Levying Tolls," *Econometrica* vol. 37, no. 1, pp. 15–24.

National Association of State Park Directors. 1984. "Annual Information Exchange," Indianapolis, Ind., Indiana Department of Natural Resources.

Nichols, D., E. Smolensky, and T. N. Tideman. 1971. "Discrimination by Waiting Time in Merit Goods," *American Economic Review* vol. 61, pp. 312–23.

Porter, Richard C. 1977. "On the Optimal Size of Underpriced Facilities," *American Economic Review* vol. 67, no. 4, pp. 753–60.

Reiling, Stephen D., Mark W. Anderson, and Kenneth C. Gibbs. 1983. "Measuring the Costs of Publicly Supplied Outdoor Recreation Facilities: A Methodological Note," *Journal of Leisure Research* vol. 15, no. 3, pp. 203–18.

Saaty, Thomas L. 1961. *Elements of Queueing Theory with Applications* (New York, Dover Publications).

Schechter, Mordechai, and Robert C. Lucas. 1978. *Simulation of Recreational Use for Park and Wilderness Management* (Baltimore, Md., Johns Hopkins University Press for Resources for the Future).

Schmanske, Stephen. 1982. "Public Goods, Product Quality Determination and Dimensionality of Consumption," *Public Finance* vol. 37, no. 3, pp. 387–403.

Shelby, Bo, and Rick Colvin. 1982. "Encounter Measures in Carrying Capacity Research: Actual, Reported and Diary Contacts," *Journal of Leisure Research* vol. 14, no. 4, pp. 350–60.

Smith, V. Kerry, ed. 1979. *Scarcity and Growth Reconsidered* (Baltimore, Md., Johns Hopkins Unviersity Press for Resources for the Future).

Smith, V. Kerry. 1981. "Congestion, Travel Cost Recreational Demand Models and Benefit Evaluation," *Journal of Environmental Economics and Management* vol. 8, pp. 92–96.

———, and Yoshiaki Kaoru, 1985. "The Hedonic Travel Cost Model: A View from the Trenches," Working Paper No. 85–W31, Department of Economics and Business Administration, Vanderbilt University, Nashville, Tennessee.

———, and J. V. Krutilla. 1976. *Structure and Properties of a Wilderness Travel Simulator* (Washington, D.C., Resources for the Future).

Stankey, George. 1973. "Visitor Perception of Wilderness Recreation Carrying Capacity," Research Paper INT-142, Forest Service, U.S. Department of Agriculture, Ogden, Utah.

U.S. General Accounting Office. 1980. "Facilities in Many National Parks and Forests Do Not Meet Health and Safety Standards" (Washington, D.C., GAO, CED-80-115).

———. 1982. "The National Park Service Has Improved Facilities at Twelve Park Service Areas," (Washington, D.C., GAO, RCED-83-65).

Walsh, Richard G., Nicole P. Miller, and Lynde O. Gilliam. 1983. "Congestion and Willingness to Pay for Expansion of Skiing Capacity," *Land Economics* vol. 59, no. 2 (May) pp. 195–210.

Weitzman, Martin L. 1974a. "Free Access vs. Private Ownership as Alternative Systems for Managing Common Property," *Journal of Economic Theory* vol. 8, pp. 225–34.

———. 1974b. "Prices vs. Quantities," *Review of Economic Studies* vol. 41, pp. 477–91.

Wilson, John D. 1983. "Optimal Road Capacity in the Presence of Unpriced Congestion," *Journal of Urban Economics* vol. 13, pp. 337–57.

Womble, P., and S. Studebaker. 1981. "Crowding in a National Park Campground: Katmai National Monument in Alaska," *Environment and Behavior* vol. 13, no. 5, pp. 557–63.

Yardas, David, A. J. Krupnick, H. M. Peskin, and W. Harrington. 1982. *Directory of Environmental Asset Data Bases and Valuation Studies* (Washington, D.C., Resources for the Future).